AF522435

ENCYCLOPAEDIC DICTIONARY OF NANOSCIENCE

ENCYCLOPAEDIC DICTIONARY OF NANOSCIENCE

[Vol. 3]

Edited by
Thomas George

ANMOL PUBLICATIONS PVT. LTD.
NEW DELHI - 110 002 (INDIA)

ANMOL PUBLICATIONS PVT. LTD.
H.O.: 4374/4B, Ansari Road, Darya Ganj,
New Delhi-110 002 (India)
Ph.: 23261597, 23278000
B.O.: No. 1015, Ist Main Road, BSK IIIrd Stage
IIIrd Phase, IIIrd Block,
Bangalore - 560 085 (India)
Visit us at: www.anmolpublications.com

Encyclopaedic Dictionary of Nanoscience

First Published, 2006

ISBN 81-261-2733-3 (Set)

PRINTED IN INDIA

Printed at Mehra Offset Press, Delhi.

Preface

This title "Encyclopedic Dictionary of Nanoscience" is one of its kinds in terms of coverage, scope, elaboration, and usefulness. Nanoscience is now a rapidly growing field of research and development that is cutting across many traditional boundaries. This technology has the ability to control or manipulate on the atomic scale. Nano is a prefix meaning one billionth (1/1,000,000,000). This is why; Nanoscience overlaps with a wide range of the existing sciences. So, whichever field of Physics, Chemistry, Computer Science or Biology (Biomedical) one specializes in, one will be able to find some degree of involvement. Degrees spread over several of these, however, have a greater bearing on Nanotechnology. Subjects that seem at present to have the most direct and constructive involvement with Nanotechnology are: Optical and Quantum Physics; Molecular Biology; Biomedical Engineering (through genetics and Microbiology); Electrical Engineering (through MEMS systems); Software Engineering (through AI, Communication and Information Systems).

Nanoscience has, therefore, become the hottest topic of discussion in the scientific world, media, government circles and corporate circles. Although it is a field that has already seen its share of hype, there is a general consensus that over the long term, nanoscience is likely to reconfigure everything. Nanoscience involves the manipulation of matter at nanometer length (one-billionth of a meter) scales to produce new materials, structures and devices. It may be defined as a technology, which involves research and technology development involving structures with at least one dimension in approximately the 1-100 nanometer range, frequently with atomic/molecular precision. It also involves processes for creating and using structures, devices and systems that have novel properties and functions because of their nanometer scale dimensions.

Nanoscience is a completely new production process in a molecular world, opening up possibilities far beyond what we have known before. Nature provides the inspiration for this new technology – from time immemorial, nanomachines have been at work in all organic cells, building our world of plants, animals and people from a store of chemical elements.

Nanotube is real. Strictly speaking, any tube with nanoscale dimensions, but generally used to refer to carbon nanotubes (a commonly mentioned non-carbon variety is made of boron nitride), which are sheets of graphite rolled up to make a tube. The dimensions are variable (down to 0.4 nm in diameter) and you can also get nanotubes within nanotubes, leading to a distinction between multi-walled and single-walled nanotubes. Apart from remarkable tensile strength, nanotubes exhibit varying electrical properties (depending on the way the graphite structure spirals around the tube, and other factors), and can be insulating, semiconducting or conducting (metallic).

Nanoelectronics is electronics on a nanometer scale, whether made by current techniques or nanotechnology; includes both molecular electronics and nanoscale devices resembling today's semiconductor devices. Experts world over believe nanotechnology will have a more far-reaching impact than the silicon integrated chip with many more applications than electronics. Nanotechnology is expected to revolutionize certain areas, such as semiconductors, pharmaceuticals and materials. The total societal impact of nanotechnology is expected to be much greater than that of the silicon integrated circuit because it is applicable in many more fields than electronics. Nanosurgery is a generic term including molecular repair and cell surgery. The nanotechnological innovation succeeded in revolutionizing the entire computer hard drive market in less than ten years. These days GMR reading heads are the key component in the global market for hard drives, estimated to be worth over US$40 billion. The hard drives in commercially available computers are now able to store 80 gigabytes of data, thanks to the nanotechnological GMR effect. Not only has the GMR effect succeeded in completely taking over the billion-dollar hard drive reading head market, it has also increased performance from 40 megabytes to the current level of 80 gigabytes. Thus, thanks to nanotechnology, computer hard drives are already two thousand times more efficient. Nanotechnological breakthroughs of this kind enable investors to participate in newly emerging billion-dollar markets within a few years.

Major areas where Nanotech products will invite some kind of regulation by government agencies include: Biological products (vaccines, blood products, tissues); Cosmetics; Devices; Foods (for humans and animal feed, though generally not meat and poultry); Dietary supplements; Drugs (human and animal); Radiation Emitting Electronic Products; Color additives used in food, drugs, cosmetics, devices; Combination products (i.e., drug-device, drug-biologic, and device-biologic products).

This encyclopedic dictionary tends to cover the aforementioned subject areas in their broadest possible terms. All useful supplementary research and reference tools have been included for readers' further investigations in the subject area of Nanotechnology. This Encyclopedia seeks to explain this complex field of study in a rather user-friendly manner. That is why the whole Encyclopedia is divided into ten volumes giving the readers a logically flowing perspective of the subject area of nanotechnology.

In sum, this Encyclopedic Dictionary seeks to provide the reader a thorough understanding of what is happening now in the field of nanotechnology, as well as what to expect in future by generations to come. This could be considered as the easiest reference book on the subject of Nanoscience. The editor believes that this Encyclopedic Dictionary will go a long way as the most worthwhile introductory book on the subject.

—Editor

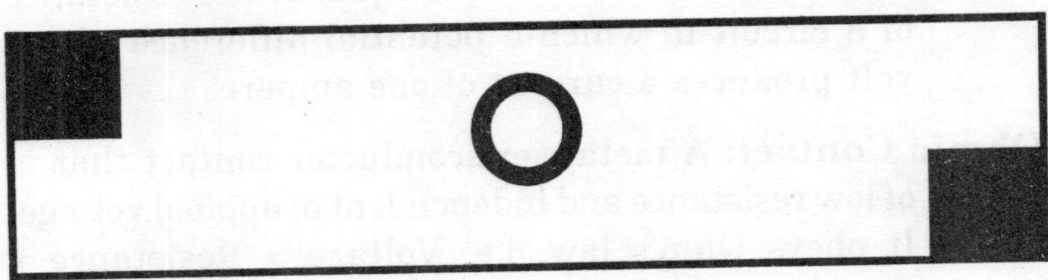

OEE (Overall Equipment Effectiveness): The actual output of a tool divided by the potential output of a tool assuming 100% utilization and 100% uptime.

OEIS OptoElectronic Integrated Systems: research is currently focused on the development of methodologies for self-assembly of micron scale objects with the objective of fabricating optoelectronic integrated systems (OEIS) [RP25]. In an extension of our integrated microarray research, electronically addressable test chips, comprising n x n matrices of microelectrodes each bearing photolithographically defined oligonucleotides of programmed base sequence, have been developed as experimental platforms for the self- assembly and interconnection of oligonucleotide modified optoelectronic components, e.g., LEDs and photodetectors. In this process, DNA- modified optoelectronic components may be transported to the surface of a microelectrode using electrophoresis whereupon sequence specific oligonucleotide interactions direct component localisation and binding.

OEMs (Original Equipment Manufacturers): Companies that: sell products after buying individual components from a variety of other manufacturers.

O-E-O: The conversion of optical to electrical to optical signals in switches that are not completely optical.

Ohm: A unit of electric resistance equal to the resistance of a circuit in which a potential difference of one volt produces a current of one ampere.

Ohmic Contact: A metal semiconductor contact that is of low resistance and independent of applied voltage. It obeys Ohm's law, i.e. Voltage = Resistance x Current.

Okazaki Fragments: Short lengths of DNA produced on the lagging strand during DNA replication, discovered by R. Okazaki. They are rapidly joined by DNA ligase to form a continuous DNA strand.

OLED or Organic LED: OLED is not made of semiconductors. It's made from carbon-based molecules. That is the key science factor that leads to potentially eliminating LEDs' biggest drawback ñ size. The carbon-based molecules are much smaller. And according to a paper written by Dr. Uwe Hoffmann, Dr. Jutta Trube and Andreas Kl–ppel, entitled OLED—A bright new idea for flat panel displays "OLED is brighter, thinner, lighter, and faster than the normal liquid crystal (LCD) display in use today. They also need less power to run, offer higher contrast, look just as bright from all viewing angles and are—potentially—a lot cheaper to produce than LCD screens." See also LCD and LED. LCD, LED, and OLED definitions courtesy The San Francisco Consulting Group (SFCG)

Oligodendrocyte: Type of glial cell in the vertebrate central nervous system that forms a myelin sheath around axons.

Oligomer: Short polymer, usually consisting (in a cell) of amino acids (oligopeptides), sugars (oligosaccharides), or nucleotides (oligonucleotides). (From Greek *oligos*, few, little.)

Olympus: At its launch on July 12,1989, the largest civilian telecommunications satellite in the world. Olympus was an experimental satellite built for the European Space Agency (ESA) by British Aerospace.

Omega Point: Also called the Quantum Omega Point Theory. A possible future state when intelligence controls the Universe totally, and the amount of information processed and stored goes asymptotically towards infinity. See Terminology From The Omega Point Theory List . [Origin: Teilhard de Chardin, The Phenomenon of Man. See also Barrow and Tipler, The Cosmological Anthropic Principle or Tipler's The Physics of Immortality for a more modern definition.] [AS]

Omitted Reaction: A chemical reaction that fails by not occurring .

Oncogene: One of a large number of genes that can help make a cell cancerous. Typically, a mutant form of a normal gene (proto-oncogene) involved in the control of cell growth or division.

One-Chip Solution: A sensor element and its electronics integrated on one chip.

Oocyte: Developing egg; usually a large and immobile cell.

Oogenesis: Formation and maturation of oocytes in the ovary.

Op-Amp (Operational Amplifier): An amplifier having

a large voltage gain, very high input impedance and low output impedance. Its output is dependent on the instantaneous differences between the two input waveforms.

Op-amp: (1) Operational Amplifier — semiconductor amplifier characterized by high gain and high internal resistance. (2) Logic gate whose output is a binary 1 if any of its inputs is a 1; zero otherwise.

Opaquing: The applying of minute droplets of opaquing ink to fill pinholes and correct geometric areas on masks or reticles.

Operation: A designer-specified process capability that is the actual product that designers order from the MEMS Exchange. Process capabilities can have a number of parameters, some of which may be specified by the designer. For instance, a deposition process would require knowledge of the thickness of the deposition, an etch knowledge of the etch depth, etc When a designer specifies values for all the parameters that require a value, an operation is created.

Operator: Short region of DNA in a bacterial chromosome that controls the transcription of an adjacent gene.

Operon: In a bacterial chromosome, a group of contiguous genes that are transcribed into a single mRNA molecule.

Optical Add/Drop Multiplexer (OADM): A device that adds or drops individual wavelengths from a DWDM system.

Optical Attenuator: A passive device for reducing the amplitude of a signal without appreciably distorting the waveform.

Optical Backplanes: A novel optoelectronic device that eliminates the need for "electrical to optical and optical to electrical" conversions and achieves 10 Gbit/s speeds and beyond, in optical telecommunications. Backplanes refer to the bus or matrix that is at the back of a modular networking product, and into which the modules are plugged.

Optical Cross-Connect: A network device used by telecom carriers to switch high-speed optical signals (OC-12, OC-48, etc.). It differs from digital cross-connects in that it deals with multiple high-speed signals that are switched in their entirety and not multiplexed together. Optical cross-connects work entirely at the optical layer and may be able to operate without having to convert to electrical form and back again.

Optical Dispersion: The cause of bandwidth limitations in a fiber. Dispersion causes the spreading or broadening of light pulses as they travel through a fiber. The three major types are modal dispersion, chromatic dispersion, and waveguide dispersion.

Optical Fabrics: An optical switch fabric enables telecom service providers to switch any type of data traffic.

Optical Interconnect Devices: The linkage used to join two or more communications units, such as systems, networks, links, nodes, equipment, circuits, and devices.

Optical Lithography: The use of lenses and light to precisely project and expose a mask pattern onto a resist-coated wafer.

Optical Proximity Correction (OPC): A photolithography procedure that introduces selective image size alterations into the reticle pattern to compensate for light diffraction and interference between closely spaced features.

Optical Pyrometer: A contactless temperature-sensing device based on measuring the infrared radiation given out by a heated body.

Optical Ring Protection: The most robust protection scheme employed to achieve optical restoration, i.e. to avoid the implications when a single fiber gets cut, affecting a large number of optical networks connected to the cut fiber. Optical ring protection requires more numbers of fibers than other restoration techniques but it is the most sophisticated.

Optical Switch: A switch that allows signals in optical fibers to be selectively switched from one circuit to another. It could either be through mechanical means or electrical and magnetic means.

Optical Trapping: Optical trapping of small particles by forces exerted by laser radiation pressure has recently been introduced into the study of biological systems. The single- beam gradient trap (optical tweezers) employs a single strongly focused laser beam. In this case, particle size is much less than and the laser light exerts a force pulling the particle toward the high focus part of the beam. Such manipulation of micron- sized (particle size larger than) particles is also feasible. Manipulations of cells and intracellular organelles have extended to laser cutting (scissors) and to use of two- photon systems. [National Center for Research Resources "Integrated Genomics Technologies Workshop Report" Jan 1999]. A method for non- mechanical manipulation of small particles. Small transparent objects (such as glass- spheres or yeast cells) can be trapped using focused laser light. The method has many possible applications in various fields of physics and biology. The optical trap is based on the concept of photon momentum transfer. Photon momentum was proposed by Planck and used by Einstein to explain the photo-electric effect.

Optical Tunneling: A quantum mechanical phenomena resulting from photon delocalization that allows light to cross propagation barriers such as an interface.

Optical Tweezers: Developed at Bell Labs in the 1980s ... The method relies on the fact that light waves exert minute forces as well as transmitting energy. If an object is small enough— in this case about 5 micrometers across — it can be "trapped" or held still by a focused beam of laser light. When the laser's position changes, the trapped object moves too ... The new technique allows researchers to make measurements not possible with conventional methods and should make it easier to judge the effectiveness of inhibitors and other medicines at a variety of concentrations. The single-beam gradient trap, or the *optical tweezers*, is a unique tool for micro- manipulation, enabling its user to access tiny objects without any mechanical contact. It uses a focussed laser beam to trap and manipulate microscopic particles in the size range 100 nm - 100 mm. The trap consists of a single laser beam which attracts transparent dielectric particles towards its focal region. The laser beam is introduced into a conventional optical microscope, so that the same objective is used to view and trap particles immersed in a liquid.

Optoceramic Materials: A class of proprietary materials having high electro-optic coefficients manufactured using chemical-vapor-deposition techniques that can be used to reduce the size of optical devices

Optoelectronics: ICs and systems, which combine optical elements (e.g. lasers and photo detectors) with microelectronic circuits, often as part of a single chip.

OR Gate: The output is "yes" if at least one of the inputs is a "yes."

Orbit: The path of a satellite, planet, or heavenly body around another, larger, body in space. For example, the Earth is in orbit around the Sun. The Moon is in orbit around the Earth. Orbits differ in their eccentricities.

Orbital Tower: Also known as a "space tether", "beanstalk" or "heavenly funicular". A cable in synchronous orbit, with one end anchored to the surface of the Earth, often with a small asteroid at the outer end to provide some extra tension and stability. Picture also a "space elevator". In theory, constructed of a diamondoid material, approximately 22,000 miles long, with one end in a stable orbit, and the other somewhere [probably] around the equator. Used frequently in science-fiction yarns, and may become a reality with the advent of mature MNT. Such an elevator would move freight and passengers into orbit at a cost per pound orders of magnitude less than current launches, with passenger safety comparable to train, plane, or subway trips. Becomes possible when we can mass-produce nanotubes, and make their length to fit.

Orbital Waste: The space near Earth is being polluted

with small orbiting projectiles, some as small as a pin. Most of the debris is floating fragments of discarded rocket stages, but it also includes gloves and cameras dropped by astronauts. This is not a problem for life on Earth, but it is a problem as life begins its historic spread beyond Earth—the first great expansion since the greening of the continents, long ago. Orbiting objects travel much faster than rifle bullets, and energy increases as the square of speed. Small fragments of debris in space can do tremendous damage to a spacecraft, and worse—their impact on an spacecraft can blast loose yet more debris. Each fragment is potentially deadly to a spacefaring human crossing its path. Today, the tiny fraction of space that is near Earth is increasingly cluttered. This litter needs to be picked up. With molecular manufacturing, it will be possible to build small spacecraft able to maneuver from orbit to orbit in space, picking up one piece of debris after another. Small spacecraft are needed, since it makes no sense to send a shuttle after a scrap of metal the size of a postage stamp. With these devices, we can clean the skies and keep them hospitable to life

Orbital: In the approximation that each electron in a molecule has a distinct, independent *wave function*, the spatial distribution of an electron wave function corresponds to a molecular orbital. These, in turn, can be approximated as sums of contributions from the orbitals characteristic of the isolated atoms. An electron added to a molecule—or, similarly, one excited to a higher-energy state within a molecule—would occupy a state with a different wave function from the rest; an unoccupied state of this kind corresponds to an unoccupied molecular orbital.

Orbital-symmetry effects on reaction rates arise when a reaction requires *overlap* between two lobes of the orbitals on each of two *reagents*: if the algebraic signs of the wave functions in the facing lobes do not match, bond formation between those orbitals is prohibited.

Organic Compound: A chemical compound of carbon, hydrogen, and one or more other elements, like oxygen or nitrogen.

Organic Molecule: A molecule containing carbon; the complex molecules in living systems are all organic molecules in this sense.

Organic Synthesis: Combinatorial libraries & synthesis Narrower terms high throughput organic synthesis

Organics: Organic superconductors are a sub-class of organic conductors that include molecular salts, polymers and pure carbon systems (including carbon nanotubes and C_{60} compounds). They may also be referred to as "molecular" superconductors. They are typically large, carbon-based molecules of 20 or more atoms, consisting of a planar organic molecule and a non-organic anion.

Organotypic: Despite their wide use, the physiological relevance of organotypic slices remains controversial. Such cultures are prepared at 5 days postnatal. Although some local circuitry remains intact, they develop subsequently in isolation from the animal and hence without plasticity due to experience. Development of synaptic connectivity and morphology might be expected to proceed differently under these conditions than in a behaving animal.

Oscillator: Circuit that produces an alternating voltage (current) when supplied by a steady (DC) energy source.

Osmosis: The flow of fluid through a semi-permeable membrane from a solution with a higher concentration to a solution with a lower concentration until there is an equal concentration of fluid on both sides of the membrane. Unlike diffusion, in osmosis, chemical equilibrium between two adjacent solutions is achieved by a flow of solvent rather than a flow of solute.

Osmotic Pressure (ð): The pressure that must be applied to a solution to stop osmosis: ð = MRT.

Outgassing: The release of absorbed gases or vapors from materials usually accelerated by heating or using vacuum.

Outmessaging: In medical nanorobotics, conveyance of information from a transmitter located inside the human body, especially from working nanodevices, to the patient or to a recipient external to the human body.

Overcoming Cytoxicity: Studies have shown that quantum dot toxicity is due to cadmium ions being released into cells. This is exacerbated by oxygen, UV exposure and the large exposed surfaces of the spherical quantum dots. To avoid these problems, surfaces are added to the quantum dots, traditionally consisting of Zinc-Sulfide compounds to create a more benign surface. Polymers are used as well, which have been shown to be stable, but the polymers themselves may kill cells. Probing the Cytotoxicity of Semiconductor Quantum Dots Austin M. Derfus, Warren C. W. Chan, et al. Nano Letters 2004, Vol. 4, No. 1, 11-18

Overflow Rinser: The flowing of de-ionized water through and around wafers, sometimes with a nitrogen bubbler to aid flow at the wafer surface.

Overlap Repulsion: A repulsive force resulting from the nonbonding *overlap* of two atoms.

Overlap: *Orbitals* lack sharply defined surfaces, declining in amplitude exponentially in their surface regions. When two orbitals are brought together, regions of substantial amplitude overlap. The resulting system can be described as two new orbitals, one formed by joining the two original orbitals without introducing a node in the *wave function*, and the other formed with a node between them. The nodeless joining reduces the energy of the electrons relative to the separate orbitals, resulting in a bonding interaction; joining with a node raises the energy, producing an antibonding interaction. If both new orbitals are occupied, antibonding forces dominate, resulting in *overlap repulsion*. *Molecular mechanics* models give an approximate description of overlap (and other) forces for a certain range of atoms and geometries.

Overlay Budget: In photolithography, the maximum allowable relative displacement between a new patterned layer and the earlier defined layer.

Overlay: In lithography it refers to the superposition of a mask's pattern on a pattern already present on a wafer.

Over-Voltage Protection: Diodes are frequently used to conduct dangerously high voltages away from sensitive devices, most commonly by being reverse-biased (non-conducting) under normal circumstances, and becoming forward-biased (conducting) when the voltage rises above its normal value. For example, diodes are used in stepper motor and relay circuits to de-energize coils rapidly without the damaging voltage spikes that would otherwise occur. Many integrated circuits also incorporate diodes on the connection pins to prevent external voltages from damaging their sensitive transistors. Specialized diodes are used to protect from over-voltages at higher power.

Oxidation: The elimination of one or more electrons from an atom, ion, or molecule. It also refers to the formation of oxides of chemical elements. The removal of one or more electrons from an atom, ion or molecule.

Oxide-Trapped Charge: A charge trapped in the bulk gate oxide far away from the oxide-semiconductor interface.

Oxygen Storage: Calculations with oxygen suggest that a diamondoid sphere ~0.1 microns in diameter should easily hold oxygen at ~1,000 atmospheres. While higher pressures are feasible, they offer declining returns. At higher pressures, the pressure-volume relationship becomes severely non-linear and the density approaches a limiting value. Other gases might also be stored if diamondoid spheres can be built, but the analysis has not been done

Oxygen: A gas used in semiconductor fabrication to oxidize silicon or to form vapor-deposited oxide.

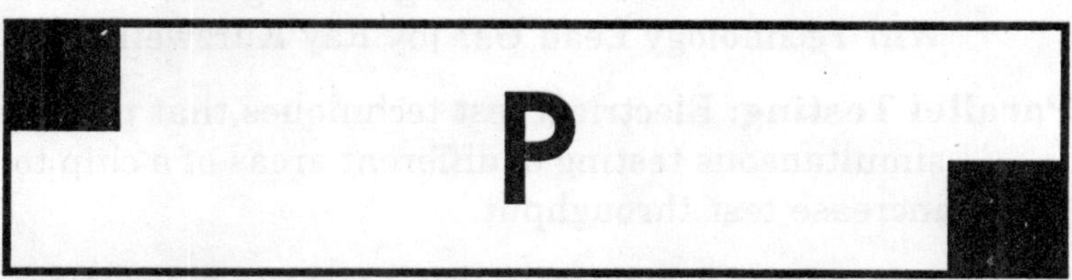

p+: A silicon region doped heavily with acceptor ions, e.g. boron, to achieve concentration levels of 10^{-20} cm.$^{-3}$

Package: Protective enclosure for a chip or a sensor, typically made of plastic or ceramic. The protective container or housing for an electronic component or die, with external terminals to provide electrical access to the components inside. Packages provide power and signal distribution, power dissipation, electrical connection and physical and chemical protection of the circuits.

Palindromic Sequence: Nucleotide sequence that is identical to its complementary strand when each is read in the same chemical direction—for example, GATC.

Paradigm Shift: When one conceptual world-view is replaced by another, or, a change of patterns on a massive scale. When Copernicus showed how the Earth rotates around the Sun, and not vice versa, that created a paradigm shift [it forced a new way of thinking about our place in the Universe]. And when quantum physics and general relativity displaced Newtonian mechanics, that created another shift. Applied to an enabling technology such as molecular manufacturing, it suggests that there will be many shifts occurring, soon, and with wide-ranging and often disruptive consequences. For more detail, see Accelerating Intelligence: Where Will Technology Lead Us? [by Ray Kurzweil].

Parallel Testing: Electrical test techniques that permit simultaneous testing of different areas of a chip to increase test throughput.

Paramagnetism: A relatively weak form of magnetism resulting from the independent alignment of atomic dipoles (magnetic) with an applied magnetic field. Also a type of induced magnetism, associated with unpaired electrons, that causes a substance to be zapped into the inducing magnetic field.

Parameter: A numerical value relating to a characteristic of a process or device.

Parametric Tests: Tests that measure DC conditions, functionality, yield, and beam characteristics of a chip or surfacemicromachined device, such as maximum current, leakage, and output drive.

Parasitic Capacitance: The undesirable, inherent capacitance of an electrical component.

Parasitic: A parasitic is an undesirable stray capacitance, inductive coupling, or resistance leakage, as well as undesired transistor actions. The first and last are most serious in monolithic ICs.

Partial Pressure of a Gas System: The pressure each gas in the system will have if all of the other gases were entirely eliminated from the system.

Particle: A very small piece of matter that moves and has energy. Particles are the fundamental things that make up the universe.

Partition Function: A function determined by the probability distribution (over *phase space* in the *classical* treatment; over *quantum* states in the quantum treatment) describing a thermally equilibrated system; many thermodynamic quantities can be expressed in terms of the partition function and its derivatives.

Passivation: Usually a silicon-dioxide or silicon-nitride layer deposited last to seal the circuit to protect it from moisture and/or contamination. The passivation layer is patterned and etched to open it up to the bonding pads.

Passive Component: An electrical component without "gain" or current-switching capability. Commonly used when referring to resistors, capacitors, or inductors.

Passive Device: Any device that does not require a source of energy for its operation. Examples include electrical resistors or capacitors, diodes, optical fibers, glass, lenses, and filters.

Patch-clamp Recording: Electrophysiological technique in which a tiny electrode tip is sealed onto a patch of cell membrane, thereby making it possible to record the flow of current through individual ion channels in the patch.

Pathogen (adjective pathogenic): An organism or other agent that causes diseases.

Pattern Generator: A system that takes CAD information and generates mask patterns. The system typically has a light source, a reduction lens system, a variable aperture, and a stepping apparatus.

Pattern Wafers: Wafers that have an image of a die

matrix formed into a layer of photoresist or a permanent pattern that has been etched into the substrate.

Patterned Etching: An etch process that uses a masking layer to define a pattern on the wafer surface to remove only selected portions of the wafer surface material.

Patterning: The passing on of an image from a reticle or mask to the photoresist coating on top of the wafer.

Payload Bay: The storage compartment of NASA's space shuttle. The Payload Bay, found on the top of the space shuttle, has large doors that open into space. When satellites are sent into space aboard the space shuttle, they are held in the Payload Bay.

Payload: The equipment carried into space by a space vehicle. The payload of a satellite includes all of the instruments and science experiments—anything that is not essential to basic functioning of the satellite, but used for science, data collection, or tools.

PCB (Printed-Circuit Board): An insulating card with one or more layers of copper interconnections onto which are soldered IC packages and other electronic components. PCBs are usually plugged into sockets that may be interconnected through a wire bus. A substrate on which a predetermined pattern of printed wiring and printed elements has been formed. Also called a printed-wiring board.

PCR Polymerase Chain Reaction: (1) High-throughput nucleic acid-based technologies are revolutionizing modern biology—none more so than PCR, the most sensitive technique to amplify and detect low abundance mRNA and minute amounts of DNA.

While PCR is the backbone of molecular biology, it has not achieved the same throughput as more recent technologies, such as microarrays, biochips, and parallel whole genome sequencing. Success in any of these areas depends on a balanced combination of good experimental design, sample preparation, probe design, amplification, detection, and analysis, as well as selection of equipment and reagents. (2) Technique for amplifying specific regions of DNA by multiple cycles of DNA polymerization, each followed by a brief heat treatment to separate complementary strands.

PECVD (Plasma Enhanced CVD): A single-wafer process where plasma is generated in the reactor to increase the energy available for the chemical reaction that deposits thin films on the wafer surface.

Pellicle: A thin, transparent membrane which seals off the mask or reticle surface from airborne contamination.

Penetration Depth (also London Penetration Depth): This term relates to how deeply a magnetic field will penetrate the surface of a superconductor. An external magnetic field impinged upon a Type 2 superconductor will decay exponentially into the surface based on the paired electron density within the superconductor (only a small fraction of the electrons are in a superconductive state). The "London" name comes from brothers F. London and H. London, who in 1935 created a theoretical model of superconductivity.

Peptide: Organic chemical found in living tissues. Peptides are polymers of amino acids. Amino acids in peptides are linked to each other by the so-called peptide (or amide) bonds, which bind the amino ($-NH_2$) group of one amino acid and the carboxyl (-COOH) group of neighboring amino acid (dehydration synthesis reaction). Some hormones are peptides. Peptides are not really different from proteins. The basic distinction is that peptides are low-weight polymers of amino acids, while proteins are polymers containing more amino acids (i.e. their molecular weight is larger—from few thousands

Peptide Bond: Chemical bond between the carbonyl group of one amino acid and the amino group of a second amino acid—a special form of amide linkage.

Peptide Map: Characteristic two-dimensional pattern (on paper or gel) formed by the separation of the mixture of peptides produced by the partial digestion of a protein.

Peptide Nanotubes: NANOTUBES formed from cyclic peptides (PEPTIDES, CYCLIC). Alternating D and L linkages create planar rings that self assemble by stacking into nanotubes. They can form pores through CELL MEMBRANE causing damage. MeSH 2004

Peptide: A short chain of amino acids; see *protein*.

Peptidomimetic: A compound containing ono- peptidic structural elements that is capable of mimicking or antagonizing the biological action (s) of a natural parent peptide. A peptidomimetic does no longer have classical peptide characteristics such as enzymatic ally scissile peptic bonds. [IUPAC Medicinal Chemistry]

Peptides are highly flexible, a quality that counteracts their receptor-affinity. Peptides are degraded by proteases in the stomach and should be administered intravenously. In the blood peptides are degraded by proteases as well and cleared from the circulation rapidly. Peptides are often too water-soluble to be able to pass the barriers that separate them from their targets in the cells and in the brain. All these disadvantages prevent peptides from becoming drugs and stimulate the quest for peptide mimics.

Percutaneous: An operation performed through a small opening in the skin.

Perimeter Array: Conventional packaging with wire bonding that has bonding pads on the die boundaries to reduce the number of input/output pins in the package.

Permeability [μ]: From the relation between magnetic induction and magnetic field (B = μH); for free space, $\mu_0 = 1.26 \times 10^{-6}$ H/m.

Permissible Exposure Limit (PEL): An exposure limit averaged over a given period or a maximum allowable exposure limit for employees.

Permitivity [å]: From the relation between polarization charge and electric field; for free space, $å_0 = 8.85 \times 10^{-12}$ F/m.

Perovskites: A large family of crystalline ceramics that derive their name from a mineral known as a perovskite. They are the most abundant minerals on earth and have a metal-to-oxygen ratio of approximately 2-to-3. Copper-oxide superconductors are layered perovskites. The perovskite name comes from Russian mineralogist Count Lev Aleksevich von Perovski.

Peroxisome: Small membrane-bounded organelle that uses molecular oxygen to oxidize organic molecules.

Contains some enzymes that produce and others that degrade hydrogen peroxide (H2O2)

Pervasive Computing: When computers (and sensors and actuators) become virtually invisible, and are used in almost every aspect of human commerce, interaction, and life. It will allow you full control over data and information, enabling you to send, receive, manage, and update your data from anywhere at any time. It will also allow you full control over your environment, in so far as you will be able to speak or gesture commands, effecting changes to things around you. Applications include: environmental monitoring—when you enter a room, they sense your presence and adjust temperature and humidity to your personal preferences; building security—to sense chemical weapons and perform face recognition; information transfer—allowing you to send and receive calls, data, and images from anywhere to anywhere, without the need of bulky equipment. Also called "Ubiquitous Computing", "Intelligent Telesensing", "Proactive Computing", "Distributed Information Management Systems", "The Evernet", and "Calm Technology". "*...it will look like nothing to the naked eye. ...beneath the surface, tiny computing networks will be doing exactly what we want them to do—working behind the scenes to help us see clearer, travel safer, and place more knowledge, rather than frustration, into our heads.*"

pH scale: A scale that specifies the strength or weakness of an acid (0 to 7) or base (7 to 14).

pH value: The negative logarithm of hydrogen-ion activity denoting the degree of acidity or baseness of a solution, at room temperature.

pH: Common measure of the acidity of a solution: "p" refers to power of 10, "H" to hydrogen. Defined as the negative logarithm of the hydrogen ion concentration in moles per liter (M). Thus pH 3 (10E-3 M H+) is acidic and pH 9 (10E-9 M H+) is alkaline.

Phage: Any virus that infects bacteria. They were the first organisms used for the study of molecular genetics and are now widely used as cloning vectors. (From Greek *phagein*, to eat.)

Phagocyte: Specific cell that occurs throughout the body and that ingests foreign matter—other cells, bacteria or nonliving matter. This process is known as phagocytosis. There are different cells that perform the task of phagocytosis in human body—granular leukocytes (a type of white blood cells) and macrophages. General term for a professional phagocytic cell—that is, a cell such as a macrophage or neutrophil that is specialized to take up particles and microorganisms by phagocytosis.

Phagocytosis: Process by which particulate material is endocytosed ("eaten") by a cell. Prominent in carnivorous cells, such as *Amoeba proteus*, and in vertebrate macrophages and neutrophils. (From Greek *phagein*, to eat.)

Pharmaceutical Profiling: The pharmaceutical properties of drug candidates determine how much of the drug safely reaches the therapeutic target. Drug candidates often fail in discovery and development due to inadequate properties, resulting in lost opportunities and resources for developing new drugs. Pharmaceutical profiling *assays* have been developed and implemented to measure the properties of large numbers of drug candidates starting at the earliest stages of discovery. This information is used for informed decisions in drug candidate selection and synthetic optimization. A holistic process of parallel activity and property optimization has emerged in drug discovery.

Pharmacodelivery: Site-directed pharmacodelivery is a desirable but elusive goal. Endothelium and epithelium create formidable barriers to endogenous molecules as well as targeted therapies *in vivo*. Deidre P. MacIntosh et. al, Targeting endothelium and its dynamic caveolae for tissue-specific transcytosis *in vivo*.

Pharmacoinformatics: PHARMINFO (Pharmacoinformatics Network) is an interdisciplinary moderated discussion forum dedicated to various aspects of implementation and use of modern information technologies in pharmacy practice, pharmacy education, *drug design* and development, and related fields. [Iosif Vaisman, "E-drug: Pharmacoinformatics" 12 Nov. 1998]

Pharmacophore Generation: A procedure to extract the most important common structural features relevant for a given biological activity from a series of molecules with a similar mechanism of action.

Pharmacyte: In medical nanorobotics, a theorized (nanorobotic) device capable of delivering precise doses of biologically active chemicals to individually-addressed human body tissue cells (e.g. cell-by-cell drug delivery).

Pharmainformatics: The multidisciplinary informatics needs of the pharmaceutical industry (HTS High Throughput Screening) data, combinatorial chemistry, ADME informatics, cheminformatics, toxicology, etc. information access and communication between various departments like the development and discovery teams.

Pharmainformatics: The multidisciplinary informatics needs of the pharmaceutical industry (HTS High Throughput Screening) data, combinatorial chemistry, ADME informatics, cheminformatics, toxicology, etc. information access and communication between various departments like the development and discovery teams.

Pharmocogenomics: The study of how an individual's genetic makeup affects the body's response to drugs. Its goal is to manufacture drugs that are tailor-made for individuals and are adapted to each person's own genetic composition.

Phase Shift: The alteration in the angular phase of a sinusoidal quantity with respect to a reference angle or another sinusoidal quantity of the same frequency. In lithography, it increases resolution of pattern transfers by destructive interference, preventing resist exposure in the regions in which it should not be exposed. It is directly proportional to the refractive index of the material. It is also defined as the lag between the input and output signals.

Phase Space: A classical system of *N* particles can be described by its 3*N* position and 3*N* momentum coordinates. The phase space associated with the system is the 6*N* dimensional space defined by these coordinates.

Phase Transformation: A change in the number and/or character of the phases that make up the microstructure of an alloy.

Phase: A portion of a material system whose properties and composition are homogeneous and which is physically distinct from other parts of the system. Any of the forms or states, solid, liquid, gas, or plasma, in which matter can exist, depending on temperature and pressure.

Phase-Shift Mask (PSM): A mask that defines regions for phase shifting to realize the shrunken featuresby changing the phase of the exposing light. It has a layer of material of a refractive index that shifts the phase of light passing through transparent portion of the mask. It improves the resolution of optical systems.

Phase-Slip (also Quantum Phase-Slip): A point where a material in a superconductive state spontaneously changes from one state to another, generating a topological "defect". This defect causes paired electrons to become "out of step" with each other, producing a voltage and, ergo, non-zero electrical resistance. This phenomenon has been observed in ultra-thin wires less than a few tens-of-nanometers in diameter. Though bulk superconductivity may persist ($T<T_c$), one consequence of phase slip is a lower current-carrying state. A similar phenomenon occurs in Josephson Junctions.

Phonon: (1) A quantum of vibration. Every harmonic vibration can be decomposed in the elementary vibrations called phonons. The total number of phonons in a system that vibrates (e.g. crystal) is related to the temperature of the system. At higher temperatures, vibration of an object is stronger and the number of phonons larger. As every phonon carries a quantum of vibrational energy, this means that the internal energy of the object is also larger.

(2) A single quantum of vibrational or elastic energy. (3) A particle of sound. The energy E of a phonon is given by the Einstein relation, $E = hf$. Here f is the frequency of the sound and h is Planck's constant. The momentum p of a photon is given by the de Broglie relation, $p = h/ë$. Here ë is the wavelength of the sound. A quantum of acoustic energy, analogous to the quantum of electromagnetic radiation, the photon. Thermal excitations in a crystal or in an elastic continuum can be described as a population of phonons (analogous to blackbody electromagnetic radiation). In highly inhomogeneous solids, a description in terms of phonons breaks down and localized vibrational modes become important.

Phosphine (PH_3): A toxic gas that is a source of phosphorus in CVD, ion implantation, and epitaxial growth.

Phosphorescence: Luminescence that occurs at times greater than on the order of a second after an electron excitation event.

Phosphorus: The n-type dopant commonly used for the buried-layer contact and emitter diffusions in standard bipolar IC technology, and the source/drain regions in NMOS.

Photo resist: A substance used to protect areas not to be etched. This process patterns via lithography followed by removal (etching) of exposed (or unexposed) areas. A light-sensitive material that is used to form selectively protective masking layers during wafer processing. A photo-mask is used to define the areas of photoresist exposed to light, which are subsequently removed by a solvent, leaving the required pattern on the wafer surface.

Photo-Acid Generator (PAG): A chemical used in deep ultraviolet (DUV) resists that boosts the resist's sensitivity to DUV light by producing a catalyst upon irradiation that weakens the linkages of the photoresist. E.g. di(tbutylphenyl) iodonium triflate.

Photobleaching: Light-induced change in a chromophore, resulting in the loss of its absorption of light of a particular wave length. The photon energy causes a conformational change in the photoreceptor proteins affecting PHOTOTRANSDUCTION. This occurs naturally in the retina (ADAPTATION, OCULAR) on long exposure to bright light. Photobleaching presents problems when occurring in PHOTODYNAMIC THERAPY, and in FLUORESCENCE MICROSCOPY. On the other hand, this phenomenon is exploited in the technique, FLUORESCENCE RECOVERY AFTER PHOTOBLEACHING, allowing measurement of the movements of proteins and LIPIDS in the CELL MEMBRANE. Cell biologists have used photobleaching to investigate the lateral mobility of fluorophores on the cell surface since the 1970s. Fusions of green fluorescent protein (GFP) to specific proteins extend photobleaching techniques to the investigation of protein dynamics within the cell, leading to renewed interest in photobleaching experiments.

Photodiode: Semiconductor diode that produces voltage (current) in response to illumination.

Photodiodes: These have wide, transparent junctions. Photons can push electrons over the junction, causing a current to flow. Photo diodes can be used as solar cells. And in photometry. If a photon doesn't have enough energy, it isn't going to turn the photodiode on very much. LED's can be used as low-efficiency photodiodes in signal applications. Sometimes a LED is paired with a photodiode or phototransistor in the same package. This device is called an "opto isolator." Unlike a transformer, this scheme allows for DC coupling. These are used to protect hospital patients from shock. Patients with IV's in their bodies are particularly susceptable, sometimes succumbing to 'carpet shock.' They are also used to isolate low-current control or signal circuitry from "dirty" power supply circuits or higher-current motor and machine circuits

Photoexcitation: Luminescence arising from photoexcitation. [IUPAC Photo]

Photolithographic Mask: A template used in photolithography that allows selective exposure of a photosensitive surface.

Photolithography: Lithographic techniques involving light as the pattern-transfer medium. See also Lithography.

Photoluminescence: Emission of light from a molecule or atom that has absorbed electromagnetic energy. Examples include fluorescence and phosphorescence.

Photo-Mask: The photographic film (glass plate) that contains the masking pattern required for a particular wafer-processing step.

Photomicrograph: The picture made with a microscope.

Photon: A quantum of electromagnetic radiation, light. Light we experience in everyday life is in fact a 'swarm' of electromagnetic quanta i.e. photons. Each photon carries a quantum of energy which is equal to Planck's constant times its frequency. Photon frequency is nontrivially related to color we observe if it is within a visible part of the electromagnetic radiation spectrum. X-ray photons for example are not visible. Due to their high energy, they can penetrate through matter.

A particle of light. The energy E of a photon is given by the Einstein relation, $E = hf$. Here f is the frequency of the light and h is Planck's constant. The momentum p of a photon is given by the de Broglie relation, $p = h/$ë. Here ë is the wavelength of the light.

Photonic Band Gap (PBG): An energy range (and corresponding wavelength range) for which a material neither absorbs light nor allows light propagation.

Photonic Band-Gap Structures: Materials with periodic changes in the dielectric constant that hinder propagation of electromagnetic waves in certain frequency ranges. They are used in lasers, antennas, millimeter-wave devices, solar cells, etc. Photonic band-gap structures can be used to trap and manipulate light, sending it down specific routes.

Photonic Crystal: Composite material made as periodical arrangement of pieces of material (e.g. spheres) embedded in a background matrix of another material that has very different (much lower) dielectric constant and/or magnetic permittivity (electromagnetic properties of material). Such structures do not allow for propagation of light of specific frequencies, due to a destructive interference of photons of such frequencies (this is a wave phenomenon also known as photonic band gap). Photonic crystals offer a possibility of manipulating the emission, propagation and splitting of electromagnetic waves which is expected to be useful for technologies of all-photon telecommunications. The production of such materials relies on nanotechnological methods (e.g. lithography, self assembly approaches...).

Photons: Very small increments or parcels of electromagnetic radiation like light.

Photoplate: The term used for a mask before images have been formed on its chrome-coated surface.

Photopolymerization: A reaction that converts monomers into polymers by the exposure of a combination of monomers to ordinary or man-made light, with or without a catalyst.

Photo-Resist Stripping: A general term that describes removal of die resist with either wet stripping or dry methods.

Phototransistor: Transistor that, when powered, produces amplified voltage (current) in response to illumination.

Physical Vapor Deposition: This is primarily a physical rather than a chemical process to deposit a material by physically dislodging atoms from the surface of a solid target. Also referred to as sputtering.

Pi bond: A *covalent bond* formed by *overlap* between two *p* orbitals on different atoms (see *sp*). Pi bonds are superimposed on *sigma bonds*, forming *double* or *triple bonds*.

Pico Technology: (trillionth of a meter) — the next step smaller, after Nano-technology. The art of manipulating materials on a quantum scale. [CA-B]

Pico: A prefix meaning a trillionth or one-millionth of one-millionth (1 x 10^{-12}). Its symbol is p.

Picoengineering: Involves engineering at the level of subatomic particles (e.g., electrons). Age of Spiritual Machines: When Computers Exceed Human Intelligence by Ray Kurzweil, Penguin paperback. *Broader terms:* microengineering, nanoengineering; Narrower term: femtoengineering

Piezoactuator: An actuator typically producing a force or displacement in response toan electrical input signal .

Piezoelectric Material: A ferroelectric material in which an electrical potential difference is created due to mechanical deformation, or conversely, in which the application of a voltage causes dimensional changes in the material.

Piezoelectric: A piezoelectric material expands or contracts when a voltage is applied across it. Piezoelectric tubes are often used to move the tip in an AFM or STM. If a piezoelectric crystal is compressed, a voltage appears across it. This effect is sometimes used to measure displacements. A commonly used piezoelectric is PZT (Pb,Zr)TiO_3.

Piezoelectric: The ability to generate a voltage when mechanical force is applied, or to produce a mechanical force when a voltage is applied to a material.

Piezoelectricity: Mechanical stress generated by an electric field/charge, and vice versa. Present in strongly anisotropic dielectrics, e.g. quartz.

Piezoresistive Effect: The change in the electrical resistivity of a metal or semiconductor that is produced by mechanical stress.

Piezoresistors: Resistors that undergo a change in resistivity when undergoing mechanical stress. See also Resistors.

PIN Diode: A PIN diode (Positive-Intrinsic-Negative diode) is a photodiode with a large, neutrally doped intrinsic region sandwiched between p-doped and n-doped semiconducting regions. A PIN diode exhibits an increase in its electrical conductivity as a function of the intensity, wavelength, and modulation rate of the incident radiation

Pin-Grid Array (PGA): Also called area array. It is a package with pins distributed over much or the entire bottom surface of the package.

Pinhole: An unwanted minute void formed in a material, especially in a metallic product such as an electroplated coating or a foil. Pinholes include defects caused by thermal contraction in oxides. The term pinhole embraces a wide variety of oxide defects and is used in a broad sense today. Listed in this category are cracks caused by thermal contraction after oxidation or by handling, and regions of oxide with low dielectric strength caused by dust particles, inadequate masking, contamination, or poor resist adhesion.

Pink Goo: (humorous) Humans (in analogy with grey goo). "Pink Goo to refer to Old Testament apes who see their purpose as being fruitful and multiplying, filling up of the cosmos with lots more such apes, unmodified." [Eric Watt Forste August 1997]

Pink Goo: (humorous) Humans (in analogy with grey goo). "Pink Goo to refer to Old Testament apes who see their purpose as being fruitful and multiplying, filling up of the cosmos with lots more such apes, unmodified."

Pinout: A representation (i.e. diagram or table) that describes the function of each pin in a chip, or each wire in a network cable.

Pipeline Problem: Innovation or Stagnation: Challenge and Opportunity on the critical path to new medical products, FDA.

Pitting: A form of very localized corrosion wherein small pits or holes form, usually in a vertical direction.

Pixel: Picture Element — smallest element of an image, such as a dot on a computer monitor screen.

pK value: A measure of the strength of an acid on a logarithmic scale. The pK value is given by log10 (1/Ka), where Ka is the acid dissociation constant pK values often are used to compare the strengths of different acids.

PLA (Programmable Logic Array): A general-purpose logic circuit containing an array of logic gates, which can be connected (programmed) to perform various functions.

Plague Insurance: Medical nanotechnologies promise to extend healthy life, but if history is any guide, they may also avert sudden massive death. The word **plague** is rarely heard today, except in relation to AIDS; it calls up visions of the Black Death of the Middle Ages, when one third of Europe died in 1346-50. A virulent influenza struck in 1918, half lost in the news of the First World War: how many of us realize that it killed at least 20 million? People often act as though plagues were gone for good, as if sanitation and antibiotics had vanquished them. But as doctors are forever telling their patients, antibiotics kill bacteria, but are useless for viruses. The flu, the common cold, herpes, and AIDS—none has a really effective treatment, because all are caused by viruses. In some African countries, as much as 10 per cent of the population is estimated to be infected with the AIDS-causing HIV virus. Without a cure soon, the steep rise in deaths from AIDS still lies in the future. AIDS stands as a grim reminder that the great plagues of history are not behind us.

Planar Devices: Semiconductor devices having planar electrodes in parallel planes, made by alternate diffusion of p- and n-type impurities into a substrate. P-type semiconductors have electron hole densities exceeding their conduction-electron densities. N-type semiconductors have conduction-electron densities exceeding their hole densities.

Planar Lightwave Circuit (PLC): A device that incorporates a waveguide fabricated in a flat material such as a thin film.

Planar Process: A fabrication process for forming IC and semiconductor components. It's based on the use of a single surface for referencing each successive operation. The planar process depends on the repeated use of silicon-dioxide (SiO_2) from one surface to control the location of impurities.

Planar: Existing essentially in a single plane; a process in which all P-N junctions intersect the top surface of the semiconductor materials, such that these intersections are permanently protected by the masking oxide, and all contacts to the device can be made to the top surface. Also refers to a process that is on a two-dimensional flat surface.

Planck's Constant: A fundamental constant of quantum physics. First employed by Max Planck in 1900 to explain the black body radiation. At that time, black body radiation could not be explained by the (classical) physics known at that time (the same problem occurred with specific heats of materials at low temperatures). Planck proposed that the energy of electromagnetic field cannot take arbitrary values, but only a discrete set of values in which each member is an integer multiple of a fundamental energy (electromagnetic field quantum, photon). Planck realized that the experimental findings can be explained by relating the energy quantum (E) to the frequency of electromagnetic field (f), i.e. Planck proposed that these should be proportional to each other, $E=h*f$. The constant of proportionality is called the Planck constant (h). It is equal to $h=6.626*10^{-34}$ Js (Joule seconds). The same approach could also explain at that time 'anomalous' specific heat of materials. In that case, the vibrations of the solid are quantized and the energy of the atoms oscillating around their equilibrium positions can take only discrete values which are integer multiples of the fundamental vibrational frequency (phonon). These later discoveries were due mainly to Einstein and Debye. This Planck's discovery marked a beginning of quantum physics, the theory explaining behavior of systems on a nanometer scale.

Plasma: An electrically conductive gas composed of ionized particles, which are used to etch unwanted material through a chemical or physical bombardment process. Plasma etching takes place in a reactor, which may be of the barrel type or the planar type.

Plasma: Gas that has very high numbers of ions (charged particles). In a plasma, the number of free electrons is almost the same as the number of positive ions.

Plasma-Induced Damage: The damage to IC devices on the wafer due to plasma during DRIE. It causes trapped charges to be formed in the gate oxide under the gate electrode, causing a breakdown of the thin gate oxide.

Plastic deformation: Permanent or nonrecoverable deformation, accompanied by permanent atomic displacement.

Plasticizer: A low molecular weight polymer additive that enhances flexibility and workability and reduces stiffness and brittleness.

PLCC (Plastic Leaded Chip Carrier): A leaded quad package—a replacement for the plastic DIP in surface-mounting organic printed-circuit boards.

PMMA (Poly Methyl Methacrylate): An electron-beam and X-ray-sensitive positive resist.

PMOS (P-channel MOSFET): A MOS device where the primary current carriers are positively charged "holes".

P-N Junction: The interface region on a doped silicon crystal in which the silicon changes from p-type to n-type silicon. A p-n junction may be created by doping adjacent regions of a semiconductor with p-type and n-type dopants. If a positive bias voltage is placed on the p-type side, the dominant positive carriers (holes) are pushed toward the junction. At the same time, the dominant negative carriers (electrons) in the n-type material are attracted toward the junction. Since there is an abundance of carriers at the junction, current can flow through the junction from a power supply, such as a battery. However, if the bias is reversed, the holes and electrons are pulled away from the junction, leaving a region of relatively non-conducting silicon which inhibits current flow. The p-n junction is the basis of an electronic device called a diode, which allows electric current to flow in only one direction. Similarly, a third region can be doped n-type or p-type to form a three-terminal device, such as the bipolar junction transistor (which can be either p-n-p or n-p-n).

$POCl_3$ (Phosphorus Oxychloride): A colorless fuming liquid used in semiconductor processing as a source of phosphorus, and an n-type dopant in silicon. Liquid $POCl_3$ is vaporized at temperatures above 106°C and transferred to a furnace where the wafer is doped.

Point Defect: A crystalline defect associated with one or several atomic sites. Defects occurring in a crystal at a specific lattice site as opposed to affecting a large area.

Poisson's Ratio: (1) For elastic deformation, the negative ratio of lateral and axial strains that result from an applied axial stress. (2) A bar of an isotropic, elastic material ordinarily shrinks laterally when it is stretched longitudinally. The lateral contracting strain divided by the applied tensile strain is Poisson's ratio, which varies from material to material.

Polar Orbit: Usually has an angle of inclination of 90 degrees to the equator. On every pass around the Earth, it passes over both the north and south poles. Therefore, as the Earth rotates to the east underneath the satellite which is travelling north and south, it can cover the entire Earth's surface.

A polar orbiting satellite covers the entire globe every 14 days.

Polar: Satellite launched on February 24, 1996 by NASA in the Global Geospace Science project. Polar is an atmospheric studies satellite in polar orbit. One purpose of Polar is to gather information that will help scientists protect future satellites from radiation and other atmospheric dangers.

Polarization (P): (1) The total electric dipole moment per unit volume of dielectric material. (2) The result of light being passed through a filter that makes the light travel in all different directions, therefore making different rays of the light behave differently from one another. The filter is called a polarizing filter.

Polarized Light: Electromagnetic radiation in which the electric component of the waves is confined to a single plane.

Polycide: Formed when a refractory metal reacts with polysilicon .It is a low-resistance material.

Polyclonal Antibodies: Antibodies produced by an animal's white blood cells (lymphocytes, specifically) in response to an antigen. This response occurs naturally or can purposely be created by injecting an animal, such as a rabbit or goat, with a specific antigen. More than one kind of anti-body is produced since more than one lymphocyte is producing antibodies. This is referred to as"polyclonal". The polyclonal antibodies are isolated from the animal and can be used for detection purposes. Because the antibodies are actually a mixture with different affinities (binding capability) for the antigen of interest, some variability in performance can occur from one test to another or one batch of antibodies to another.

Polycrystalline Si: Silicon that is composed of aggregates of individual Si crystals.

Polycyclic: A *cyclic* structure contains rings of bonds; a structure having many such rings is termed polycyclic. In the polycyclic structures of interest in this volume, a large fraction of the atoms are members of multiple small rings, resulting in considerable rigidity.

Polyimide: A compound characterized by more than one inorganic compound, which is derived from ammonia by replacement of an atom of hydrogen with another element (like a metal). A family of polymers involving carbon and nitrogen bonds, known for good thermal stability.

Polyimides: Polymeric compounds containing multiple amides, i.e., NH (nitrogen-hydrogen) groups derived from ammonia (NH_3) by the replacement of two hydrogen atoms by metals or equivalents of acid groups. Used in semiconductor processing as a dielectric and a protective coating layer on wafers since it has excellent thermal stability and a low coefficient of thermal expansion.

Polymer: A large molecule consisting of mutually chemically linked smaller molecules called monomers. For example, peptides are small polymers of amino acids (which could be called monomers in this case).

Polymer Nanocomposites: There has been enormous interest in the commercialization of nanocomposites for a variety of applications, and several of these applications will be successful in the near future. Mineral fillers, metals, and fibers have been added to thermoplastics and thermosets for decades to form composites. Compared to neat resins, these composites have a number of improved properties including tensile strength, heat distortion temperature, and modulus. Thus, for structural applications, composites have become very popular and are sold in billion pound quantities. These filled thermoplastics are sold in even larger volumes than neat thermoplastics. Furthermore, the volume of fillers sold is roughly equal to the volume of thermoplastic resin sold. Clearly, the idea of adding fillers to thermoplastics and thermosets to improve properties, and in some cases decrease costs, has been very successful for many years. More recently, with advances in synthetic techniques and the ability to readily characterize materials on an atomic scale has lead to interest in nanometer-size materials. Since nanometer-size grains, fibers and plates have dramatically increased surface area compared to their conventional-size materials, the chemistry of these nanosized materials is altered compared to conventional materials.

Polymer nanocomposites combine these two concepts, i.e., composites and nanometer-size materials. Thermoplastics filled with nanometer-size materials have different properties than thermoplastics filled with conventional materials. Some of the properties of nanocomposites, such as increased tensile strength, may be achieved by using higher conventional filler loading at the expense of increased weight and decreased gloss. Other properties of nanocomposites such as clarity or improved barrier properties cannot be duplicated by filled resins at any loading. Polymer nanocomposites were developed in the late 1980s in both commercial research organizations and academic laboratories. The first company to commercialize these nanocomposites was Toyota, which used nanocomposite parts in one of its popular car models for several years. Following Toyota's lead, a number of other companies also began investigating nanocomposites. Most commercial interest in nanocomposites has focused on thermoplastics. Thermoplastics can be broken into two groups: less expensive commodity resins and more expensive (and higher performance) engineering resins. One of the goals of nanocomposites was to allow substitution of more expensive engineering resins with a less expensive commodity resin nanocomposite. Substituting a nanocomposite commodity resin with equivalent performance as a more expensive engineering resin should yield overall cost savings.

By a strict definition of nanocomposites, i.e., any filler submicron in size, there already are significant volumes of nanocomposites being produced. These amount to more than 20 million pounds. However, since these fillers are on the upper end of the nanocomposite size range, most sources have excluded them from consideration. This report covers these products but does not add their volumes into aggregate numbers of nanocomposite markets or production volumes to comply with the established convention. At this point in time, there has been much less commercial interest in thermoset nanocomposites compared to thermoplastics. This neglect may not continue much longer since thermoset nanocomposites have some distinct advantages over neat thermoset resins. This report covers some possible markets for thermoset nanocomposites.

Polymer: A chain of numerous monomers. For example, cellulose is a polymer that is formed from the monomer glucose. A molecule consisting of many smaller sub-units covalently linked together.

Polymerase Chain Reaction (PCR): A chemical reaction that uses the polymerase enzyme to carry out *in vitro* replication of DNA. An incubation process for creating billions copies of a DNA fragment. Applications include DNA fingerprinting and genetic research.

Polymeric Materials: Materials composed of large molecules, generally based on carbon, that have been formed from the chemical bonding of smaller units (monomers). Commonly known as plastics.

Polymerization: A chemical reaction resulting in the bonding together of small molecular units (monomers) to form a much larger molecule (polymer). There are many different routes, or reaction mechanisms, for polymerization.

Polymers-biomedical: More and more therapeutic problems are relevant to the use of polymer- based therapeutic aids for a limited period of time, namely the healing time related to the outstanding capacity of living systems to self-repair, e.g. bone fracture fixation with screws and plates, of wound closure by sutures and also of drug delivery from implants or similar systems based on polymeric matrices, or on aqueous dispersions or solutions of polymers. After healing the remaining prosthetic materials or devices become foreign residues or wastes that have to be eliminated from the body. Nowadays, biocompatible polymers that can degrade in the body are developed. The degradation and the elimination of degradation by-products depend on rather complex phenomena that are presently reflected inconsistently by terms issued from the tradition because each domain has developed its own terminology almost independently. This is a source of misunderstandings, confusions and misperceptions among scientists, surgeons, pharmacists, journalists and politicians, the situation being increased by the introduction of degradable polymers in plastic waste management and environmental protection.

Polymers-biomedical: More and more therapeutic problems are relevant to the use of polymer- based therapeutic aids for a limited period of time, namely the healing time related to the outstanding capacity of living systems to self-repair, e.g. bone fracture fixation with screws and plates, of wound closure by sutures and also of drug delivery from implants or similar systems based on polymeric matrices, or on aqueous dispersions or solutions of polymers. After healing the remaining prosthetic materials or devices become foreign residues or wastes that have to be eliminated from the body. Nowadays, biocompatible polymers that can degrade in the body are developed. The degradation and the elimination of degradation by-products depend on rather complex phenomena that are presently reflected inconsistently by terms issued from the tradition because each domain has developed its own terminology almost independently. This is a source of misunderstandings, confusions and misperceptions among scientists, surgeons, pharmacists, journalists and politicians, the situation being increased by the introduction of degradable polymers in plastic waste management and environmental protection.

Polymorphisms: Variations in the DNA sequence that may be associated with certain genetic diseases.

Polysilicon: Silicon consisting of crystalline grains and used as the structural material for many MEMS devices. Sometimes called "poly." The form of silicon made of many small randomly oriented crystals. Doped poly is a conductor of electricity and is often used as an alternative to metal in interconnecting devices on ICs. Polysilicon is often deposited via CVD or PVD. Polycrystalline silicon used as conductor in integrated circuits, and especially FETs. Short for Polycrystalline Silicon, used in the manufacture of computer chips.

Polyvinyl Alcohol Emulsion: A solution made by dissolving specific amounts of a fine white synthetic powder known as polyvinyl alcohol and ammonium or potassium bichromate in specific volumes of water.

Positional Assembly: Constructing materials an atom or molecule at a time Ralph Merkle, Molecular Manufacturing, Adding Positional Control to Chemical Synthesis, Zyvex.

Positional Controlled Chemical Synthesis or Positional Synthesis: Control of chemical reactions by precisely positioning the reactive molecules, the basic principle of assemblers. [NTN]

Positional Navigation: In medical nanorobotics, a form of nanorobotic navigation in which nanodevices know their exact location inside the human body to ~micron accuracy continuously at all times.

Positional Synthesis: Control of chemical reactions by precisely positioning the reactive molecules; the basic principle of assemblers.

Positive Resist: Photoresist that is removed in areas that were not protected from exposure by the opaque regions of a mask-while remaining in regions that were protected by the mask. The positive image of the mask remains following the development.

Positive Sum: A term used to describe a situation in which one or more entities can gain without other entities suffering an equal loss—for example, a growing economy.

POSS Nanotechnology: Short for Polyhedral Oligomeric Silsesquioxanes Nanotechnology. POSS nanomaterials are attractive for missile and satellite launch rocket applications because they offer effective protection from collisions with space debris and the extreme thermal environments of deep space and atmospheric re-entry. Another application of POSS nanotechnology under development is a new high-temperature lubricant. This new nanolubricant is effective at temperatures up to 500*f*F, which is 100*f*F greater than conventional lubricants. From Technologies developed by the Propulsion Directorate's Polymer Working Group at Edwards AFB

Post Monetary Economy: After the advent of mature Nanotechnology, it is likely that our economic reality will change, possibly to the extent of eliminating currency as we know it today. See "Proposal for an Ideal Nano-Specie: Gold-Pressed Latinum" Robert A. Freitas Jr.

Posthuman: Persons of unprecedented physical, intellectual, and psychological capacity, self-programming, self-constituting, potentially immortal, unlimited individuals. [Max More]

Potential Energy Surface: The *potential energy* of a *ground-state* molecular system containing N atoms is a function of its geometry, defined by $3N$ spatial coordinates (a *configuration space*). If the energy is imagined as corresponding to a height in a $3N + 1$ dimensional space, the resulting landscape of hills, hollows, and valleys is the potential energy surface.

Potential Energy: Energy stored in an object due to its position or internal structure. The energy associated with a configuration of particles, as distinct from their motions. In macroscopic experience, potential energy can be increased (for example) by stretching a spring or by lifting a mass against a gravitational force; in molecular systems, potential energy can be increased (for example) by stretching a bond or by separating molecules against a van der Waals attraction.

Potential Therapeutics: A new branch of medical science dealing with the application of remedies to diseases based on protein identification.

Potential well: In a *potential energy surface*, the region surrounding a local energy minimum. Typically taken to include at least those points in *configuration space* such that a path of steadily declining energy can be found that leads to the minimum in question, and such that no similar path can be found to any other minimum. If the PES were a landscape, this would be the region around the minimum that could be filled with water without any flowing down and away toward another minimum.

Potentiometric Device: Monitors the voltage between a sensing electrode and a reference electrode. A high input impedance voltmeter is used to minimize current flow. The voltage typically is proportional to the logarithm of the analyte concentration.

Powder Lubrication: Initiatives to accommodate next-generation engine technologies are being pursued by engineers to prepare for the larger thrust-to-weight ratios. Hydrodynamic oils can have higher densities than powders and consequently reside in heavier packaging. They also require additional power for pumps and do not lubricate in rotating devices during start-up. Additionally, automobiles are being produced that no longer depend on gasoline as its sole source of power. Automotive companies are developing environmentally friendly vehicles, known as "green cárs" but they still rely on conventional liquid lubricants which yield pollution. At temperatures greater than 500C, conventional liquid lubricants are non-existent, hence, the advent of solid/powder lubrication. Powder lubrication is an acceptable mechanism for lubrication in high-speed tribological applications, in extreme-condition environments (e.g., space). As the next-generation of space vehicles and stations are developed, high-performance turbine engines will need solid lubricated bearing technology to replace that of conventional liquid lubricant bearing systems. Dry powder lubricants can flow like liquid lubricants, but unlike liquids, they do have some degree of ability to withstand imposed shear forces without deforming or flowing.

Power [W]: Product of voltage and current in a component; also, refers to the field of electric energy supply. The rate at which work is done, expressed as the amount of work per unit time and commonly measured in units like watts. In electronics, it is the product of voltage and current in a component.

Power Conversion: A diode is called a half wave rectifier when it is used to convert alternating current electricity into direct current, by removing the negative portion of the current. A special arrangement of four diodes that will transform an alternating current into a direct current, using both positive and negative excursions of a single phase alternating current, is known as a diode bridge, single-phase bridge rectifier, or simply a full wave rectifier. With a split (center-tapped) alternating current supply it is possible to obtain full wave rectification with only two diodes. Often diodes come in pairs, as double diodes in the same housing. When it is desired to rectify three phase power, one could rectify each of the three phases with the arrangement of four diodes used in single phase, which would require a total of 12 diodes. However, due to redundancy, only six diodes are needed to make a three phase full wave rectifier. Most devices that generate alternating current (such devices are called alternators) generate three phase alternating current. Disassembled automobile alternator, showing the six diodes that comprise a full-wave three phase bridge rectifier. For example, an automobile alternator has six diodes inside it to function as a full wave rectifier for battery charge applications. Many of the small wind turbines use three double diodes bolted to the same heatsink

Preamorphization: Implantation of a wafer with an electrically inactive species, like silicon (Si), to amorphize a thin layer of the wafer to reduce channeling during ion implantation. It also reduces transient-enhanced diffusion and lower-leakage junctions can be made

Precision: The closeness of the agreement between the results of successive measurements of the same measurand carried out subject to the same conditions. See Repeatability. The degree of reproducibility among several independent measurements of the same true value under specified conditions.

Preclinical: Drug approvals glossary What model best addresses the information being pursued in the preclinical development process? How will access to representative in vivo animal models of disease and of specific pathways improve the preclinical development process and toxicity studies? Selecting appropriate assays that will enable researchers to accurately extrapolate results to humans will give considerable leverage to the projects underway. Successful translation of biomarkers from preclinical to clinical can be enabled through the use of imaging reagents, which provide insight into the activity of a compound within a physiologically complex environment, and provide a measure of predicting success in humans. By integrating information from animals and humans we can better understand function and predict drug activity and performance. The use of biomarkers in late stage drug development will improve the decision making process and success rates.

Predeposition: The first part of a two-part diffusion. In this part, a high concentration is diffused shallowly into the surface. This acts as a source for the second or drive-in portion of the process.

Preforms, Solder: Formed pieces of solder material for closing microcircuit packages. Additionally used in some applications to attach a die or chip to a package.

Presentation Semaphore: In medical nanorobotics, a mechanical device used to display specific antigens, chemical ligands, or other molecular objects to the external environment, with the purpose of selectively modifying the chemical or other surface characteristics of a nanorobot exterior.

Pressure Sensitive: (1) a tacky adhesive that is applied to a sheet material to enable the sheet to adhere to an unrelated surface by contact and light pressure without the use of water or solvent; (2) a material that has pressure sensitive adhesive applied either at the factory or in the screen printing plant; (3) a combination of a face material, pressure sensitive adhesive, and liner paper.

Preventive Maintenance: Maintenance on equipment to prevent future random failures of the equipment or cleaning to improve the quality of the equipment yield. It is performed on a scheduled basis, based on elapsed time, the number of runs, or the number of wafers processed.

Primed *in situ* labeling: A technique that labels specific sequences in whole chromosomes by *in situ* DNA chain elongation or **PCR (polymerase chain reaction)**. [MeSH, 1999]

Principal Component Analysis (PCA): A mathematical procedure that transforms a number of correlated variables into a smaller number of uncorrelated variables. Used in electronic nose technology as a method to analyze the data obtained from the sensors used.

Printed Circuit Board (PCB): Selectively metallized insulating sheet for supporting and interconnecting circuit components.

Printed-Wiring Board (PWB): A platform upon which electronic components such as IC chips and capacitors are mounted. A PWB also provides electrical interconnection between components. It consists of a fiberglass substrate on which electrical circuits are built. A bare PCB, i.e. a PCB without any components, is a PWB.

Privileged Structures: The retrospective analysis of the chemical structure of various drugs used in medicine led medicinal chemists to identify some molecular motifs that are more frequently associated with higher biological activity than other structures. Such molecular motifs were called "privileged" structures by Evans et al., to mean substructures that confer activity to two or more different receptors. The implication was that the privileged structure provides the scaffold, and that substitutions on the scaffold provide specificity for a particular receptor. Two monographs describe the privileged structure concept. Historically, the most popular privileged structures were the aryl- ethylamines (including indolyl- ethylamines), the diphenylmethane derivatives, the tricyclic psychotropics, and the sulfonamides. The dihydropyridines, the benzodiazepines, the N-arylpiperazines, and the biphenyls are more recent contributions.

Probability Density Function (pdf): A mathematical expression describing the functional relationship between a specific value of an attribute or variable and the probability of obtaining that value. Consider an uncertain physical property and a corresponding space describing the range of values that the property can have (e.g., the configuration of a thermally excited N particle system and the corresponding $3N$ dimensional configuration space). The probability density function associated with a property is defined over the corresponding space; its value at a particular point is the probability per unit volume that the property has a value in an infinitesimal region around that point.

Probability Distribution: A list of probabilities, which can have a finite number of values, associated with each of a variable's possible values. It is also called the probability function or the probability mass function.

Probability: The ratio of a number of outcomes corresponding to an event to the total number of outcomes. For example, the probability of an even number occurring when a dice is thrown is 3/6 = 0.5

Probe Card: The interface between an automatic tester and the device being tested. It is usually a printed-circuit board that has probes to make electrical contact to the device being tested.

Probing: A term used to describe testing techniques of individual IC dies by using very finely tipped probes to temporarily connect each to a test computer, thus verifying proper operation. A bad die will usually be marked with a spot of ink.

Process Capability: A comparison of the actual variability of a process to the process specification.

Process Design Rules: Specifications that ensure that size, spacing, and overlap of geometry are correct for a manufacturing process.

Process Hierarchy: A tree that organizes abstract descriptions of everything that can be done to a semiconductor wafer in transforming it into a collection of finished devices including inspection and testing. Briefly, one node in the process hierarchy might be "deposition/CVD/LPCVD," meaning that "LPCVD" is a kind of "CVD," which in turn is a kind of "deposition."

Process Integration: The optimization of each process step to be compatible with previous and succeeding steps in a chronological process flow.

Process Latitude: The ability of a process that has varying parameters to manufacture products that meet specifications. A robust process must have large process latitude to allow for variations in its parameters and ensure correct operation.

Process Module: A bundle of commonly associated processing services (such as base processes) made available by a particular fab in a set sequence, often for a special price. Every process module is identified by a generic process or a location in the process hierarchy. For example, a "deposition" module might include a cleaning, the actual deposition, and an inspection.

Product Definition: A set of models, drawings, specifications, and other information that describes a product and product's performance specifications and defines product identification elements. The product definition may include additional design and production requirements to ensure that the required design is achieved.

PROM (Programmable Read-Only Memory): A read-only memory, which can be programmed after manufacture by external equipment. Typically, PROMs utilize fuseable links, which may be burned open to produce a logic bit in a specific location.

Proof Mass: A predetermined test mass in a measuring device or machine, such as in acceleration measurement equipment, which serves as the reference mass for the quantity to be measured.

Propellant Gas: A compressed inert gas (such as a fluorocarbon) that acts as a vehicle for discharging the contents of an aerosol container.

Protein: High molecular weight (1000-1000000) amino acid polymer. Basically, the protein can be described as a sequence (chain) of amino acids. However, due to the forces acting between the amino acids, the protein structure is not chain-like, the protein folds into a complex shape. Nevertheless, long parts of the proteins are nearly linear. In these parts (also called strands or polypeptide chains) amino acids are arranged in helical, more or less straight chains. It is not difficult to imagine that extremely large number of proteins can be created from twenty amino acids. It is estimated that human body contains about 30000 different proteins.

Protein Chips: Arrays that consist of proteins themselves (e.g., for studies of protein, protein interactions or protein/small-molecule binding) or of probes for capturing proteins (so that protein levels in a sample can be gauged).

Protein Design, Protein Engineering: The design and construction of new proteins; an enabling technology for nanotechnology. [FS]

Protein Folding: "The process by which proteins acquire their functional, preordained, three-dimensional structure after they emerge, as linear polymers of amino acids, from the ribosome." [The Scientist]

Protein: (1) High molecular weight (1000-1000000) amino acid polymer. Basically, the protein can be described as a sequence (chain) of amino acids. However, due to the forces acting between the amino acids, the protein structure is not chain-like, the protein folds into a complex shape. Nevertheless, long parts of the proteins are nearly linear. In these parts (also called strands or polypeptide chains) amino acids are arranged in helical, more or less straight chains. It is not difficult to imagine that extremely large number of proteins can be created from twenty amino acids. It is estimated that human body contains about 30000 different proteins. (2) Living cells contain many molecules that consist of amino acid polymers folded to form more-or-less definite three-dimensional structures; these are termed proteins. Short polymers lacking definite three-dimensional structures are termed peptides. Many proteins incorporate structures other than amino acids, either as covalently attached side chains or as bound ligands. Molecular objects made of protein form much of the molecular machinery of living cells.

Proteomics: The study of the full expression of proteins in an organism The term proteome refers to all the proteins expressed by a genome, and thus proteomics involves the identification of proteins in the body and the determination of their role in physiological and pathophysiological functions. ... Ultimately it is believed that through proteomics new disease markers and drug targets can be identified that will help design products to prevent, diagnose and treat disease. The term proteome refers to all the proteins expressed by a genome, and thus proteomics involves the identification of proteins in the body and the determination of their role in physiological and pathophysiological functions. ... Ultimately it is believed that through proteomics new disease markers and drug targets can be identified that will help design products to prevent, diagnose and treat disease. [e-proteomics.net]

Proximal Probes: A family of devices capable of fine positional control and sensing, including scanning tunneling and atomic force microscopes; an enabling technology for nanotechnology.

Proximity Aligner: An optical system that uses a mask that is very close to the wafer.

Proximity Effect: The phenomenon where a thin film of non-superconductive material in close proximity with a superconductor takes on superconductive properties. The Josephson junction is a device that takes advantage of this phenomenon. The Inverse Proximity Effect is where just the opposite occurs. A non-superconductive metal can enhance the Tc of an adjacent superconductor. This inverse effect has been observed with silver and lead.

Proximity Printing: The exposure of a wafer by passing light through a mask that is very close to, but not in contact with, the photoresist-coated wafer. Collimated light is required to ensure minimum image resolution.

PSTN (Public Switched Telephone Network): The worldwide interconnected public-telephone network, both private and government-owned.

P-type Doping : The purpose of p-type doping is to create an abundance of holes. In the case of silicon a trivalent atom, such as boron, is substituted into the crystal lattice. The result is that an electron is missing from one of the four possible covalent bonds. Thus the atom can accept an electron from the valence band to complete the fourth bond, resulting in the formation of a hole. Such dopants are called acceptors. When a sufficiently large number of acceptors are added, the holes greatly outnumber the excited electrons. Thus, the holes are the majority carriers, while electrons are the minority carriers in p-type materials. Blue diamonds (Type IIb), which contain boron impurities, are an example of a naturally occurring p-type semiconductor

P-Type Semiconductor: A semiconductor crystal containing a small amount of "dopant" atoms that have one less outer electron than the other atoms. Each "dopant" atom causes one unoccupied spot, called a "hole," among the electrons that are bound in their orbits. The holes are positively charged and can move to constitute electric current. Boron is a commonly used P-type dopant for silicon.

Puddle Develop: A resist development process that applies a small puddle of photoresist to form a meniscus over the entire wafer to minimize required developer volume.

Punchthrough: A condition that occurs when depletion regions of the drain and source meet, causing a channel current to flow irrespective of gate voltage, which causes device failure. Punchthrough occurs due to the presence of small channel length and high electric fields at the drain end of the channel. Thinner oxides, larger substrate doping, shallower junctions and longer channels can circumvent this.

Purge: A cleaning procedure that uses an inert gas to remove undesirable gases from a process chamber.

Puzzle Chains: The advice of the tourguide, you grab two molecular knobs on the protein and pull. It resists for a moment, but then a loop comes free, letting other loops flop around more, and the whole structure seems to melt into a writhing coil. After a bit of pulling and wrestling, the protein's structure becomes obvious: It is a long chain–longer than you are tall, if you could get it straight—and each segment of the chain has one of several kinds of knobs sticking off to the side. With the multicolored, glassy-bead portrayal of atoms, the protein chain resembles a flamboyant necklace. This may be decorative, but how does it all go back together? The chain flops and twists and thrashes, and you pull and push and twist, but the original tight, solid packing is lost. There are more ways to go wrong in folding up the chain than there are in solving Rubik's Cube, and now that the folded structure is gone, it isn't even clear what the result should look like. How did those twentieth-century researchers ever solve the notorious "protein folding problem"? It's a matter of record that they started building protein objects in the late 1980s. This protein molecule won't go back together, so you try to break it. A firm grip and a powerful yank straightens a section a bit, but the chain holds together and snaps back. Though unfolding it was easy, even muscles with the strength of steel—the strength of Superman—can't break the chain itself. Chemical bonds are amazingly strong, so it's time to cheat again. When you say, "Flimsy world–one second!" while pulling, your hands easily move apart, splitting the chain in two before its strength returns to normal. You've forced a chemical change, but there must be easier ways since chemists do their work without tiny superhands. While you compare the broken ends, they thrash around and bump together. The third time this happens, the chain rejoins, as strong as before. This is like having snap-together parts, but the snaps are far stronger than welded steel. Modern assembler chemistry usually uses other approaches, but seeing this happen makes the idea of molecular assembly more understandable: Put the right pieces together in the right positions, and they snap together to make a bigger structure. Remembering the "Whoa!" command, you decide to go back to the properly scaled speed for your size and strength. Saying "Standard settings!," you see the thrashing of the protein chain speed up to hard-to-follow blur.

PVD (Plasma Vapor Deposition): A SEMICON process used to deposit material onto a wafer by ionizing gas plasma (typically via microwave radiation). In PVD, chemical reactions occur in the gas/plasma phase, as opposed to on the surface of the wafer in CVD.

P-Wave: A rare form of electron pairing in which two electrons travel together in spherical orbits; with both having the same direction of rotation.

Pyrex: Also called borosilicate glass. A type of glass resistant to heat, chemicals, and electricity. Used in chemical apparatus, industrial equipment, and thermometers.

Pyroelectricity: The property of certain crystals, such as tourmaline, of acquiring opposite electrical charges on opposite faces when heated.

Pyrolytic: A chemical reaction induced by external heat.

Pyrophoric: Materials that ignite in air at or below 54.4^{0}C.

PZT (Lead Zirconate Titanate): A family of ceramics that has low electrical losses and high values for the piezoelectric charge coefficient, electromechanical coupling coefficient, and dielectric constant. Used extensively in ultrasound transducers.

Q Factor (Quality Factor): A measure of the sharpness of resonance of an electronic/electro-mechanical resonator component. A device with a high Q factor has a sharp, large magnitude, and well-defined resonance. It is defined as the ratio of the reactance to the effective series resistance of a component. A rating, applied to coils, capacitors, and resonant circuits, equal to the reactance divided by the resistance. The ratio of energy stored to energy dissipated per cycle in an electrical or mechanical system.

QC-1 (Quality Criteria-1): The basic quality principles and requirements for nuclear weapons research, design, development, procurement, production, dismantlement, maintenance, stockpile evaluation, and disassembly/disposal.

QFP (Quad Flat Pack): A common package for ICs with high pin count. It has many leads on all 4 sides of the package. QFP is losing popularity in the IC industry.

Quadrature: A phenomenon when two periodic signals of the same frequency are out of phase by 90°(360°/4).

Quadrupole Analyzer: An analyzer in which, during each short time interval, ions of a particular mass are allowed to pass through the analyzer and are counted by the detector. This is repeated for other types of ions that have different masses. Thus the analyzer scans through a large range of masses. Used in electronic noses

Quadrupole Mass Analyzer (QMA): A mass analyzer that separates the ions generated in the ion source according to their mass-to-charge ratio and is controlled by a combination of RF and DC potentials. Only ions of a specific mass can pass through a given set of oscillating RF and DC fields developed through four cylindrical rods.

Qualification Plan (PQ): The qualification plan documents the engineering evaluation objectives, the engineering evaluation results, and performance metrics for prescribed attributes. A control number for callout in an application shall identify each PQ.

Qualification: The process of assuring that a product and all associated processes are capable of meeting customer requirements.

Quantization: The concept that energy can occur only in discrete units called quanta.

Quantum Biology: Quantum physics and molecular biology are two disciplines that have evolved relatively independently. However, recently a wealth of evidence has demonstrated the importance of quantum mechanics for biological systems and thus a new field of quantum biology is emerging. Living systems have mastered the making and breaking of chemical bonds, which are quantum mechanical phenomena. Absorbance of frequency specific radiation (e.g. photosynthesis and vision), conversion of chemical energy into mechanical motion (e.g. ATP cleavage) and single electron transfers through biological polymers (e.g. DNA or proteins) are all quantum mechanical effects. Hopefully, the merging of disciplines known as nanotechnology will remove the interface between quantum physics and biology. In a paper titled 'The Importance of Quantum Decoherence in Brain Processes,' Max Tegmark sought to prove that the brain is too warm to maintain the coherence required for quantum computation. From the results of his calculations, Tegmark claims that, "there is nothing fundamentally wrong with the current classical approach to neural network simulations." This statement contradicts the hypothesis that the brain functions as a quantum computer, originally proposed by Roger Penrose. Tegmark's claim was amplified by a recent report in *Science* beginning with the sentence, 'Sir Roger Penrose is incoherent, and Max Tegmark says he can prove it.' However, the computations carried out by Tegmark relied on a value of 310K for the temperature in his model of the neuron. While the average kinetic energy (temperature) of an entire brain cell may be 310K, the most fundamental characteristic of life is that it is not at equilibrium and thus, our statistical method for measuring temperature breaks down at small sizes, especially at the nanoscale. Biological systems are known to have ways of manipulating local temperatures. For instance, Koichiro Matsuno has determined by the postulate of black body radiation measurements that actomyosin complexes (abundant in the axons of nerve cells) can reach local temperatures as low as $1.6*10^{-3}$K. Matsuno argues that actomyosin functions as a heat engine (a device that converts heat energy into mechanical energy) that is able to maintain a constant velocity due to quantum mechanical coherence and entanglement.

Quantum Computer: (1) A computer that exploits quantum mechanical phenomena such as superposition and entanglement. (2) Computation using the systems whose behavior is importantly determined by the laws of quantum mechanics. The systems in question are called quantum bits or qubits, and can be e.g. nuclei, semiconductor quantum dots and similar. The most important property that these systems must have is that they can support the coherence (phase) of the wavefunction of their quantum excitation, i.e. they should be isolated for the classical macroscopic surrounding (heat bath) which decoheres the quantum excitation in a qubit. Quantum computing is based on the principle of superposition of states which applies to quantum-mechanical systems. Simply speaking, the quantum system can be simultaneously in both zero (0) and one (1) states, i.e. the state of the quantum system is in general superposition of *off* (0) and *on* (1) states. This is in contrast to classical bits which are either in on state *or* in off state. Possibly the most famous illustration of the superposition principle is the well known Schrodinger cat paradox. It is known that the computation with qubits can speed up certain mathematical tasks enormously (e.g. decomposition of a number in its prime factors). However, the number of algorithms for which the quantum computation is the preferred choice is small (only two or three). Some envision quantum computers that use solid-state type of qubits (quantum dots), i.e. a precisely nanostructured material that can be considered as an array of qubits, but the production of uniform nanoscale arrays of qubits that are sufficiently isolated from the classical, decohering surrounding may prove to be a quite demanding technological task.

Quantum Computing: Computation using the systems whose behavior is importantly determined by the laws of quantum mechanics. The systems in question are called quantum bits or qubits, and can be e.g. nuclei, semiconductor quantum dots and similar. The most important property that these systems must have is that they can support the coherence (phase) of the wavefunction of their quantum excitation, i.e. they should be isolated for the classical macroscopic surrounding (heat bath) which decoheres the quantum excitation in a qubit. Quantum computing is based on the principle of superposition of states which applies to quantum-mechanical systems. Simply speaking, the quantum system can be simultaneously in both zero (0) and one (1) states, i.e. the state of the quantum system is in general superposition of *off* (0) and *on* (1) states. This is in contrast to classical bits which are either in on state *or* in off state. Possibly the most famous illustration of the superposition principle is the well known Schrodinger cat paradox. It is known that the computation with qubits can speed up certain mathematical tasks enormously (e.g. decomposition of a number in its prime factors). However, the number of algorithms for which the quantum computation is the preferred choice is small (only two or three). Some envision quantum computers that use solid-state type of qubits (quantum dots), i.e. a precisely nanostructured material that can be considered as an array of qubits, but the production of uniform nanoscale arrays of qubits that are sufficiently isolated from the classical, decohering surrounding may prove to be a quite demanding technological task.

Quantum Confined Atoms (QCA): atoms caged inside nanocrystals. May find uses in clear-glass sunglasses, bio-sensors, and optical computing.

Quantum Corral: Structures where electrons are corralled (confined) by atoms placed on the surface of a metal. See Don Eiger's STM gallery for examples.

Quantum Cryptography: A system based on quantum-mechanical principles. Eavesdroppers alter the quantum state of the system and so are detected. Developed by Brassard and Bennett, only small laboratory demonstrations have been made. [AS]

Quantum dot (Qdots): Nanometer sized semiconductor particles, made of cadmium selenide (CdSe), cadmium sulfide (CdS) or cadmium telluride (CdTe) with an inert polymer coating. The semiconductor material used for the core is chosen based upon the emission wavelength range being targeted: CdS for UV-blue, CdSe for the bulk of the visible spectrum, CdTe for the far red and near-infrared, with the particle's size determining the exact color of a given quantum dot. The polymer coating safeguards cells from cadmium toxicity but also affords the opportunity to attach any variety targeting molecules, including monoclonal antibodies directed to tumor-specific biomarkers. Because of their small size, quantum dots can function as cell- and even molecule-specific markers that will not interfere with the normal workings of a cell. In addition, the availability of quantum dots of different colors provides a powerful tool for following the actions of multiple cells and molecules simultaneously. In

August 2004, researchers announced the successful preparation of water-soluble gold quantum dots that can also be constructed to emit light at a variety of wavelengths. These polymer-coated quantum dots may prove to be more suitable for use in human clinical applications.

Sometimes called single-electron-transistor, quantum bit, or qubit; particle of matter so small that the addition of a single electron (single electron transfer) profoundly changes its properties. The attribute quantum serves to remind that the behavior of electron in such structures must be described in terms of quantum theory. Atoms are examples of quantum dots. Structures made of few to few hundreds of atoms are also quantum dots (e.g. cadmium selenide or gallium arsenide nanocrystals, clusters), thus their characteristic scale is nanometer.

The presence or absence of a single electron in such structures can be used to store information (that's the reason for their alternative name—qubits). Thus, their possible use is in nanoscale electronics. Features on nanostructured surfaces (e.g. corals) can also serve as quantum dots. Quantum dots are nanometer-scale "boxes" for selectively holding or releasing electrons. Over the past decade or so they have been transformed from laboratory curiosities to the building blocks for a future computer industry. Quantum dots are small devices that contain a tiny droplet of free electrons.

They are fabricated in semiconductor materials and have typical dimensions between nanometres to a few microns. The size and shape of these structures and therefore the number of electrons they contain, can be precisely controlled; a quantum dot can have anything from a single electron to a collection of several thousands.

Atomic physics progressed rapidly at the beginning of the last century, thanks, in large part, to optical spectroscopy. Quantization and spin were discovered through optical studies, as were other fundamental atomic properties. With the advent of the laser, physicists learned how to manipulate atomic wave functions by applying coherent optical fields. More discoveries followed. Now, at the beginning of the new century, optical techniques are being used to explore a new scientific frontier: the atom-like entities known as quantum dots (QDs).

Measuring 1-100 nm across, QDs are semiconductor structures in which the electron wave function is confined in all three dimensions by the potential energy barriers that form the QD's boundaries. A QD's electronic response, like that of a single atom, is manifest in its discrete energy spectrum, which appears when electron hole pairs are excited. Although the wave function of a QD electron, and its corresponding hole, extends over many thousands of lattice atoms, the pair— termed an 'exciton'—behaves in a quantized and coherent fashion. The coherence is relatively easy to detect and control optically—for two reasons. First, the superposition of the ground and excited states dephases more slowly in QDs than in higher-dimensional semiconductor structures. Second, QDs have large dipole moments (50-100 times larger than those of atoms). Thanks to these advantages, it is possible to probe and manipulate the wave function of a single QD. QDs possess another attractive property. Their size, shape, and composition can all be tailored to create a variety of desired properties. These "artificial atoms" can, in turn, be positioned and assembled into complexes that serve as new materials. Researchers who work on QDs anticipate that a host of complex, customized QD-based materials will become available. The physics of quantum dots shows many parallels with the behaviour of naturally occurring quantum systems in atomic and nuclear physics. As in an atom, the energy levels in a quantum dot become quantized due to the confinement of electrons. Unlike atoms however, quantum dots can be easily connected to electrodes and are therefore excellent tools to study atomic-like properties. There is a wealth of interesting phenomena that have been measured in quantum dot structures over the past decade. This page shows a few examples from our group. The next paragraph first discusses briefly the parallels between atoms and quantum dots.

Quantum Dot Lasers: Semiconductor lasers are key components in a host of widely used technological products, including compact disk players and laser printers, and they will play critical roles in optical communication schemes. The basis of laser operation depends on the creation of nonequilibrium populations of electrons and holes, and coupling of electrons and holes to an optical field, which will stimulate radiative emission. Calculations carried out in the early 1970s by C. Henry predicted the advantages of using quantum wells as the active layer in such lasers: the carrier confinement and nature of the electronic density of states should result in more efficient devices operating at lower threshold currents than lasers with "bulk" active layers. In addition, the use of a quantum well, with discrete transition energy levels dependent on the quantum well dimensions (thickness), provides a means of "tuning" the resulting wavelength of the material. The critical feature size—in this case, the thickness of the quantum well—depends on the desired spacing between energy levels. For energy levels of greater than a few tens of millielectron volts (meV, to be compared with room temperature thermal energy of 25 meV), the critical dimension is approximately a few hundred angstroms. Although the first quantum well laser, demonstrated in 1975, was many times less efficient than a conventional laser, the situation was reversed by 1981 through the use of new materials growth capabilities (molecular beam epitaxy), and optimization of the heterostructure laser design. Even greater benefits have been predicted for lasers with quantum dot active layers. Arakawa and Sakaki predicted in the early 1980s that quantum dot lasers should exhibit performance that is less temperature dependent than existing semiconductor lasers, and that will in particular not degrade at elevated temperatures. Other benefits of quantum dot active layers include further reduction in threshold currents and an increase in differential gain—that is, more efficient laser operation. Some of the key concepts in the laser operation. Stimulated recombination of electron-hole pairs takes place in the GaAs quantum well region, where the confinement of carriers and of the optical mode enhance the interaction between carriers and radiation. In particular, note the change in the electronic density of states, as a function of the "dimensionality" of the active layer. The population inversion (creation of electrons and holes) necessary for lasing occurs more efficiently as the active layer material is scaled down from bulk (3-dimensional) to quantum dots

(0-dimensional). However, the advantages in operation depend not only on the absolute size of the nanostructures in the active region, but also on the uniformity of size. A broad distribution of sizes "smears" the density of states, producing behavior similar to that of bulk material. Thus, the challenge in realizing quantum dot lasers with operation superior to that shown by quantum well lasers is that of forming high quality, uniform quantum dots in the active layer. Initially, the most widely followed approach to forming quantum dots was through electron beam lithography of suitably small featured patterns (~300 Å) and subsequent dryetch transfer of dots into the substrate material. The problem that plagued these quantum dot arrays was their exceedingly low optical efficiency: high surface-to-volume ratios of these nanostructures and associated high surface recombination rates, together with damage introduced during the fabrication itself, precluded the successful formation of a quantum dot laser

Quantum Dot Nanocrystals (QDNs): Used to tag biological molecules, and "measuring between five and ten nanometres across, are made up of three components. Their cores contain paired clusters of atoms such as cadmium and selenium that combine to create a semiconductor. This releases light of a specific colour when stimulated by ultraviolet of a wide range of frequencies. These clusters are surrounded by a shell made of an inorganic substance, to protect them. The whole thing is then coated with an organic surface, to allow the attachment of proteins or DNA molecules. By varying the number of atoms in the core, QDNs can be made to emit light of different colours." [From The Economist print edition]

Quantum Dot: An object so small that adding or removing a single electron represents a significant change. Nanometer sized fragments (the dots) of semiconductor crystalline material which emits PHOTONS. The wavelength is based on the quantum confinement size of the dot. They are brighter and more persistent than organic chemical INDICATORS. They can be embedded in MICROBEADS for high throughput ANALYTICAL CHEMISTRY. Do not confuse with microscopic fluorescent bar codes which are micrometer sized. An important strategy for nonisotopic labeling of *single molecules* is the use of highly luminescent semiconductor *nanocrystals*, or 'quantum dots,' that can be covalently linked to biological molecules. This class of detectors, which range in size from 1-5 nm, have been exploited for biological labeling by a number of laboratories, particularly those of Shimon Weiss, Paul Alivisatos and Shuming Nie (4, 5). Quantum dots offer several advantages over organic *dyes*, including increased brightness, stability against photobleaching, a broad continuous excitation spectrum, and a narrow, tunable, symmetric emission spectrum. Because quantum dots are nontoxic and can be made to dissolve in water, efforts are underway to explore their use in labeling single molecules in living cells. Nanometer-sized semiconductor crystals, or electrostatically confined electrons. Something (usually a semiconductor island) capable of confining a single electron, or a few, and in which the electrons occupy discrete energy states just as they would in an atom (quantum dots have been called "artificial atoms"). Other terminology reflects the preoccupations of different branches of research: microelectronics folks may refer to a "single-electron transistor" or "controlled potential barrier," whereas quantum physicists may speak of a "Coulomb island" or "zero-dimensional gas" and chemists may speak of a "colloidal nanoparticle" or "semiconductor nanocrystal." All of these terms are, at various times, used interchangeably with "quantum dot," and they refer more or less to the same thing: a trap that confines electrons in all three dimensions.

Sometimes called single-electron-transistor, quantum bit, or qubit; particle of matter so small that the addition of a single electron (single electron transfer) profoundly changes its properties. The attribute quantum serves to remind that the behavior of electron in such structures must be described in terms of quantum theory. Atoms are examples of quantum dots. Structures made of few to few hundreds of atoms are also quantum dots (e.g. cadmium selenide or gallium arsenide nanocrystals, clusters), thus their characteristic scale is nanometer. The presence or absence of a single electron in such structures can be used to store information (that's the reason for their alternative name—qubits). Thus, their possible use is in nanoscale electronics. Features on nanostructured surfaces (e.g. corals) can also serve as quantum dots.

Nanometer sized fragments (the dots) of semiconductor crystalline material which emits PHOTONS. The wavelength is based on the quantum confinement size of the dot. They are brighter and more persistent than organic chemical INDICATORS. They can be embedded in MICROBEADS for high throughput ANALYTICAL CHEMISTRY. Do not confuse with microscopic fluorescent bar codes which are micrometer sized. MeSH 2004. An important strategy for nonisotopic labeling of single molecules is the use of highly luminescent semiconductor nanocrystals, or 'quantum dots,' that can be covalently linked to biological molecules. This class of detectors, which range in size from 1- 5 nm, have been exploited for biological labeling by a number of laboratories, particularly those of Shimon Weiss, Paul Alivisatos and Shuming Nie (4, 5). Quantum dots offer several advantages over organic dyes, including increased brightness, stability against photobleaching, a broad continuous excitation spectrum, and a narrow, tunable, symmetric emission spectrum. Because quantum dots are nontoxic and can be made to dissolve in water, efforts are underway to explore their use in labeling single molecules in living cells.

Nanometer sized fragments (the dots) of semiconductor crystalline material which emits PHOTONS. The wavelength is based on the quantum confinement size of the dot. They are brighter and more persistent than organic chemical INDICATORS. They can be embedded in MICROBEADS for high throughput ANALYTICAL CHEMISTRY. Do not confuse with microscopic fluorescent bar codes which are micrometer sized. MeSH 2004. An important strategy for nonisotopic labeling of single molecules is the use of highly luminescent semiconductor nanocrystals, or 'quantum dots,' that can be covalently linked to biological molecules. This class of detectors, which range in size from 1- 5 nm, have been exploited for biological labeling by a number of laboratories, particularly those of Shimon Weiss, Paul Alivisatos and Shuming Nie (4, 5). Quantum dots offer several advantages over organic dyes, including increased brightness, stability against photobleaching, a broad continuous excitation spectrum, and a narrow, tunable, symmetric emission spectrum. Because quantum dots are nontoxic and can be made to dissolve in water, efforts are underway to explore their use in labeling single molecules in living cells.

Quantum Interferometric Lithography: A method of lithography that exploits quantum entanglement in order to focus a lithographic beam beyond half of its wavelength.

Quantum Mechanics: A largely computational physical theory explaining the behavior of quantum phenomena, which incorporates the theory of special relativity. Despite dilignet attempts, general relativity has not been sucessfully incorporated into quantum mechanics. A physical model of chemical and optical phenomena, as well as the behaviour of matter in general on a small scale. Quantum mechanics describes a system of particles in terms of a wave function defined over the configuration space of the system. Although the concept of particles having distinct locations is implicit in the potential energy function that determines the wave function (e.g., of a ground-state system), the observable dynamics of the system cannot be described in terms of the motion of such particles from point to point. In describing the energies, distributions, and behaviors of electrons in nanometer-scale structures, quantum mechanical methods are necessary. Electron wave functions help determine the potential energy surface of a molecular system, which in turn is the basis for classical descriptions of molecular motion. Nanomechanical systems can almost always be described in terms of classical mechanics, with occasional quantum mechanical corrections applied within the framework of a classical model. A physical model of chemical and optical phenomena, as well as the behaviour of matter in general on a small scale. Quantum mechanics describes a system of particles in terms of a *wave function* defined over the *configuration space* of the system. Although the concept of particles having distinct locations is implicit in the *potential energy* function that determines the wave function (e.g., of a *ground-state* system), the observable dynamics of the system cannot be described in terms of the motion of such particles from point to point. In describing the energies, distributions, and behaviors of electrons in nanometer-scale structures, quantum mechanical methods are necessary. Electron *wave functions* help determine the *potential energy surface* of a molecular system, which in turn is the basis for *classical* descriptions of molecular motion. *Nanomechanical* systems can almost always be described in terms of classical mechanics, with occasional quantum mechanical corrections applied within the framework of a classical model.

Quantum Mirage: A nanoscale property that may allow information to be transfered through use of the wave property of electrons. Thus, quantum computers might not require wires as we know them. [NTN]

Quantum Number: A property of quantum system. Set of quantum numbers that describe the system can be used to uniquely characterise the system. Quantum system can be in different states which can be enumerated by the different sets of quantum numbers.

Quantum Physics: A branch of physics based on quantum

theory. The laws of classical physics do not hold at the scales of interest to nanotechnology. Thus, all the phenomena occurring on a scale of atoms must be described using formalism of quantum physics. The objects on a scale of nanometer do not behave as classical bodies which follow trajectories under influence of forces. Rather, they behave both as waves and rigid bodies, depending on a particular situation and a measurement one performs on them. This comes from the fact that in quantum theory energy can be emitted or absorbed by matter only in discrete units called *quanta* (singular *quantum*).

Describes fundamental electronic and optical properties of matter at microscopic level and wave and interference phenomena in particular. Quantum electronics, quantum optics and optoelectronics are important areas of application—atomic clocks, lasers, light emitting diodes, optical fibers, tunnel diodes and superconducting systems are important and well-known examples. The trend towards faster and more complicated microprocessors and microelectronics has resulted in electronic components now approaching the domains of quantum physics at research level. In about 20 years time, miniaturization will also be halted in commercial applications. It will then probably utilize electronic wave phenomena and single- electron effects in semiconductor components and systems and also to create more complicated transistor components connected in more complicated ways where quantum phenomena, cooperative phenomena and even superconductivity may be important for function. "QNANO"—*quantum nanophysics*—will probably provide a target for research and development in semiconductor physics, molecular electronics and bioelectronics in the foreseeable future.

Quantum Teleportation: 'Teleportation' (or copying) of the quantum state from one particle to another. Not really spectacular as in Star Trek serial where macroscopic objects (humans, weapons and similar) are 'teleported'. Present technology can 'teleport' quantum objects such as photons, electrons ans similar. It is important to note that the object itself is not teleported during this process. What actually happens is that the quantum state (e.g. spin number) is exactly copied from one object to another, both spatially and temporally present *prior* to 'teleportation'. The technique used is based on the Einstein-Podolsky-Rosen effect (EPR) or paradox, which relies on the so-called nonlocality of the wavefunction. An EPR source produces pair of entangled particles, one of which serves as a 'carbon copy'. The third particle (original, source particle to be copied, introduced besides the EPR pair) is 'entangled' with one of the particles from the EPR pair which enables the copying of its state to another EPR particle that has never interacted with the original, yet it becomes its copy. The experimental setup requires one measuring apparatus that entangles the original with one of the EPR particles and another apparatus that acts on another EPR particle which becomes the copy. The experimentalists operating the two apparatuses must be in contact i.e. the one that acts on the copy particle must know the results of the entanglement measurement on the source (original) particle and another EPR particle which is 'waisted' in the process. It should also be noted here that none of the experimentalists does not know the quantum state of the original particle, even when the whole process finishes, neither is it measured during the process.

Quantum Tunneling: When electrons pass through a barrier, without overcoming it or breaking it down. See Quantum tunneling and this illustration

Quantum Well: A P-N-P junction in which the "N" layer is ~10 nm (where traditional physics leaves off and quantum effects take over) and an "electron trap" is created. "If one makes a heterostructure with sufficiently thin layers, quantum interference effects begin to appear prominently in the motion of the electrons. The simplest structure in which these may be observed is a quantum well, which simply consists of a thin layer of a narrower-gap semiconductor between thicker layers of a wider-gap material." See Center for Quantum Electronics U of Dallas

Quantum Wire: Another form of quantum dot, but unlike the single-dimension "dot," a quantum wire is confined only in two dimensions—that is it has "length," and allows the electrons to propagate in a "particle-like" fashion. Constructed typically on a semiconductor base, and (among other things) used to produce very intense laser beams, switchable up to multi-gigahertz per second.

Quantum: Describes a system of particles in terms of a wave function defined over the configuration of particles having distinct locations is implicit in the potential energy function that determines the wave function, the observable dynamics of the motion of such particles from point to point. In describing the energies, distributions and behaviours of electrons in nanometer-scale structures, quantum mechanical methods are necessary. Electron wave functions help determine the potential energy surface of a molecular system, which in turn is the basis for classical descriptions of molecular motion. Nanomechanical systems can almost always be described in terms of classical mechanics, with occasional quantum mechanical corrections applied within the framework of a classical model.

QuantumBrain: [theoretical] Think of your brain. Now, think of your brain performing at vastly superior levels. Nanobots will become an as-needed addition to your existing neurons, extending your mental capacities further then you can probably now imagine. [uhf]

Quartz (Crystalline SiO_2): An inert, highly stable, and piezoelectric material.

Quartz Micro Balance Sensor Module: An array of quartz oscilloscopes, each layered with different polymers (polysiloxanes, etc.). Volatile molecules are adsorbed by the sensitive coating, which results in nanometric changes in coating mass and frequency. Best suited for detecting large molecules.

Quasar: Puts out tremendous amounts of energy from a very small source. Known as the most intense concentrations of energy in the entire universe, quasars shine with the light of a hundred galaxies, but they are no larger than our solar system.

Quasiparticle: A bare particle that is "dressed" or "clothed" by a cloud of other surrounding particles. Quasiparticles behave similarly to bare (normal) particles, but usually have a larger effective mass due to this cloud moderating interactions with other particles.

Quaternary Structure: Three-dimensional relationship

of the different polypeptide chains in a protein complex.

Qubit: The quantum computing analog to a bit. Qubits exhibit superposition. Thus, unlike normal bits, qubits can be both 1 and 0 at the same time.

Quench: The phenomenon where superconductivity in a material is suppressed; usually by exceeding the maximum current the material can conduct (Jc) or the maximum magnetic field it can withstand (Hc).

Quencher: A molecular entity that deactivates (quenches) an excited state of another molecular entity, either by energy transfer, electron transfer, or by a chemical mechanism. [IUPAC Photo]

Quenching: (1) Arresting the course of a chemical reaction by chemical or physical means. (in photochemistry) (2) The deactivation of an excited molecular entity intermolecularly by an external environmental influence (such as a quencher) or intramolecularly by a substituent through a nonradiative process. (3) (in radiation chemistry) The process of inhibiting continuous or multiple discharges following a single event in certain types of radiation detectors.

Queue Time: The highest amount of time allowed between two sequential operations.

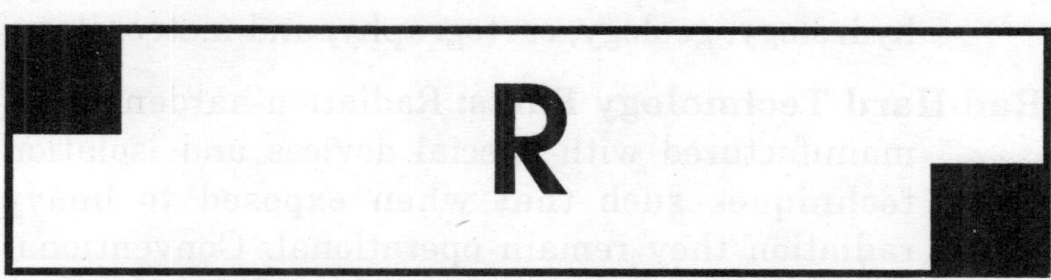

R&D Discovery: Across the entire division, R&D has been mandated to be more efficient, lower costs and generate more product. Discovery as a whole, is extremely complex. Operationally, there are large multifunction teams, decentralized processes, short-term thinking and lack of strategic information. Technically, there are silos, various methodologies, budget constraints and an abundance of technological options from which to choose. As the organization continues to demand more output and productivity from discovery, the need to manage all of the interconnected steps increases. The early stages of discovery are the hardest to manage from a technical and operational standpoint. Expectantly plagued with failures, there is a constant need for resource shifts, technological enhancements and strategic changes ... juggling priorities, measuring strategies and managing budgets, makes synthesizing these an intimidating and difficult task.

Radar: By emitting high frequency radio waves and measuring where and how fast they were reflected, a radar instrument can measure things like distance, direction, speed, etc. A radar instrument can "see" objects in the dark as well as penetrate cloud cover.

Radarsat: A revolutionary new satellite launched by the Canadian Space Agency on November 4, 1995. This remote sensing satellite uses radar to take images useful for agriculture, oceanography, forestry, hydrology, geology, cartography, and meteorology.

Rad-Hard Technology Parts: Radiation-hardened ICs manufactured with special devices and isolation techniques such that when exposed to heavy radiation they remain operational. Conventional circuits will short out under such conditions because the radiation generates electrical currents inside the semiconductor material.

Radiation Damage: Chemical changes (bond breakage, ionization) caused by high-energy radiation (e.g., x-rays, gamma rays, high-speed electrons, protons, etc.).

Radiation: The harmful combination of waves and particles, known as high energy rays, emitted from a source.

Radical: A reactive atom or group of atoms created when an energetic electron bombards a neutral gas molecule.

Radical: A structure with an unpaired electron (but excluding certain metal ions). In organic molecules, a radical is often associated with a highly reactive site of reduced valence (see *doublet*). The term radical is sometimes used to describe a substructure within a molecule; the term *free radical* then describes a radical in this sense, viewed as the result of cleaving the bond linking the substructure to the rest of the molecule.

Radio Telescope: An instrument to pick up radio waves for analysis. Radio telescopes "see" things that are too far away to see in normal light by usimg radio waves. Satellites often use parabolic dishes with radio telescopes.

Radioactive Isotope: Form of an atom with an unstable nucleus that emits radiation as it decays.

Radioactive Waste: Currently more than 100,000 tonnes of radioactive waste is stored in theUK, awaiting final disposal. The majority of the existing UK nuclear power stations will need to be decommissioned in the next 20-30 years, which will generate a substantial amount of radioactive waste to be disposed. The UK currently has no long-term solution. In 2002 the Government formed the Committee on Radioactive Waste Management (CoRWM) to make recommendations by the end of 2006 on the best option or combination of options in managing the UKs solid radioactive waste. The Royal Society has a long standing interest in the scientific aspects of this problem.

Radiometer: An instrument to measure the changing levels of radiation as well as visible and infrared light. They can produce cloud images even at night.

RAM (Random-Access Memory): The basic (read/write) storage element in a computer. It stores digital information temporarily and can be changed as required.

RAM: Random Access Memory — read-write memory with elements accessible in any order.

Random Error: A component of the error of measurement, which in the course of a number of measurements of the same measurand varies in an unpredictable way. It is not possible to correct for random error.

Random Numbers: Numbers that have no order or any

relation to each other. Sophisticated computer programs generate random numbers.

Range (Ion Implantation): The depth an ion travels in the silicon substrate during ion implantation.

Range: The difference between the minimum and maximum values of sensor output in the intended operating range. Defines the overall operating limits of a sensor.

Rapid Prototyper: A machine that can manufacture objects directly (usually, though not necessarily, in plastic) under the control of a computer.

Rapid Thermal Processing (RTP): A process that rapidly increases the temperature of a wafer by placing it in a heated oven for a very short period of time. A halogen quartz lamp that provides the heat and temperatures during RTP can reach as high as 800°C.

Ras Protein: One of a large family of GTP-binding proteins that help relay signals from cell-surface receptors to the nucleus. Named for the *ras* gene, first identified in viruses that cause rat sarcomas.

Rational Drug Design: The input of biocomputing in drug discovery is twofold: firstly the computer may help to optimise the pharmacological profile of existing drugs by guiding the synthesis of new and "better" compounds. Secondly, as more and more structural information on possible protein targets and their biochemical role in the cell becomes available, completely new therapeutic concepts can be developed. The computer helps in both steps: to find out about possible biological functions of a protein by comparing its amino acid sequence to databases of proteins with known function, and to understand the molecular workings of a given protein structure. Understanding the biological or biochemical mechanism of a disease then often suggests the types of molecules needed for new drugs

RCA Clean: A wet cleaning technique that has three distinct steps: removal of organic contaminants from the wafer surface using a 5:1:1 H_2O:H_2O_2 :NH_4OH solution; removal of any oxide layer that may have built up using a 50:1 H_2O:HF solution; and removal of ionic or heavy metal contaminants using a solution of 6:1:1 H_2O:H_2O_2 :HCl. Named after the now defunct Radio Corporation of America, the company for which the inventor of the RCA clean formula worked.

Reactance: Portion of impedance that characterizes non-dissipative, energy storage effects (also see "impedance"). The component of the impedance of an AC circuit that is due to capacitance or inductance or both. Its units are ohms.

Reactant: (1) A starting substance in a chemical reaction. It appears to the left of the arrow in a chemical reaction. (2) A substance that is consumed in the course of a chemical reaction. It is sometimes known, especially in the older literature, as a reagent, but this term is better used in a more specialized sense as a test substance that is added to a system in order to bring about a reaction or to see whether a reaction occurs (e.g. an analytical reagent).

Reaction: (1) A process that transforms one or more chemical species into others. Typical reactions make or break bonds; others change the state of ionization or other properties taken to distinguish chemical species. (2) In chemistry, any process in which the arrangement of atoms into molecules is changed.

Reaction-Rate Limited: During CVD at lower pressures, the arrival rate of reactants surpasses the rate at which they are consumed by the chemical reaction process at the wafer surface.

Reactivity: The ability of a chemical to undergo chemical reactions and release energy.

Reactor: A process chamber in which chemical reactions occur to deposit specific materials in layers during semiconductor manufacturing.

Reading Frame: The phase in which nucleotides are read in sets of three to encode a protein; an mRNA molecule can be read in any one of three reading frames.

Reagent Device: A large *reagent* structure (or a large structure that binds a smaller reagent) serving as a component of a *mechanochemical* system. A reagent device exists chiefly to hold, position, and manipulate the environment of a *reagent moiety*.

Reagent Moiety: The portion of a reagent device that is intimately involved in a chemical reaction.

Reagent: A chemical *species* that undergoes change as a result of a chemical *reaction*.

Reagents: Substances used for the detection, identification, analysis, etc. of chemical, biological, or pathologic processes or conditions. .. Reagents are substances used for the detection or determination of another substance by chemical or microscopical means, especially analysis. Types of reagents are precipitants, solvents, oxidizers, reducers, fluxes, and colorimetric reagents.

Rebuilding Tissues: Again, skin provides easy examples and may be a natural place to start in practice.

People often want hair where they have bare skin, and bare skin where they have hair. Cell herding machines could move or destroy hair follicle cells to eliminate an unwanted hair, or grow more of the needed cells and arrange them into a working follicle where a hair is desired. By adjusting the size of the follicle and the properties of some of the cells, hairs could be made coarser, or finer, or straighter, or curlier. All these changes would involve no pain, toxic chemicals, or stench. Cell-herding devices could move down into the living layers of skin, removing unwanted cells, stimulating the growth of new cells, narrowing unnaturally prominent blood vessels, insuring good circulation by guiding the growth of any needed normal blood vessels, and moving cells and fibers around so as to eliminate even deep wrinkles. At the opposite end of the spectrum, cell herding will revolutionize treatment of life-threatening conditions. For example, the most common cause of heart disease is reduced or interrupted supply of blood to the heart muscle. In pumping oxygenated blood to the rest of the body, the heart diverts a portion for its own use though the coronary arteries. When these blood vessels become constricted, we speak of coronary-artery disease. When they are blocked, causing heart muscle tissue to die, we speak of someone "having a coronary," another term for heart attack. Devices working in the bloodstream could nibble away at atherosclerotic deposits, widening the affected blood vessels. Cell herding devices could restore artery walls and artery linings to health, by ensuring that the right cells and supporting structures are in the right places. This would prevent most heart attacks. But what if a heart attack has already destroyed muscle tissue, leaving the patient with a scarred, damaged, and poorly functioning heart? Once again, cell-herding devices could accomplish repairs, working their way into the scar tissue and removing it bit by bit, replacing it with fresh muscle fiber. If need be, this new fiber can be grown by applying a series of internal molecular stimuli to selected heart muscle cells to "remind" them of the instructions for growth that they used decades earlier during embryonic development. Cell-herding capabilities should also be able to deal with the various forms of arthritis. Where this is due to attacks from the body's own immune system, the cells producing the damaging antibodies can be identified and eliminated. Then a cell-herding system would work inside the joint where it would remove diseased tissues, calcified spurs, and so forth, then rework patterns of cells and intercellular material to form a healthy, smoothly working, and pain-free joint. Clearly, learning to repair hearts and learning to repair joints will have some basic technologies in common, but much of the research and development will have to be devoted to specific tissues and specific circumstances. A similar process—but again, specially adapted to the circumstances at hand—could be used to strengthen and reshape bone, correcting osteoporosis. In dentistry, this sort of process could be used to fill cavities, not with amalgam, but with natural dentin and enamel. Reversing the ravages of periodontal disease will someday be straightforward, with nanomedical devices to clean pockets, join tissues, and guide regrowth. Even missing teeth could be regrown, with enough control over cell behavior

Receptor: (1) A structure that can capture a molecule (often of a specific type in a specific orientation)

owing to complementary surface shapes, charge distributions, and so forth, without forming a covalent bond. See dissociation constant. Protein that binds a specific extracellular signaling molecule (ligand) and initiates a response in the cell. Cell-surface receptors, such as the acetylcholine receptor and the insulin receptor, are located in the plasma membrane, with their ligand-binding site exposed to the external medium. Intracellular receptors, such as steroid hormone receptors, bind ligands that diffuse into the cell across the plasma membrane. (2) Protein that binds a specific extracellular signaling molecule (ligand) and initiates a response in the cell. Cell-surface receptors, such as the acetylcholine receptor and the insulin receptor, are located in the plasma membrane, with their ligand-binding site exposed to the external medium. Intracellular receptors, such as steroid hormone receptors, bind ligands that diffuse into the cell across the plasma membrane.

Receptor-Based Transport: The most efficient of all modes of molecular transport involves receptor sites capable of recognizing and selectively binding specific molecular species. Many receptors reliably bind only a single molecular type; others, such as sugar transporters, can recognize and transport several related sugar molecule types. In nanomechanical systems, artificial binding sites of almost arbitrary size, shape, and electronic charge may be created and employed in the construction of a variety of highly efficient molecular sortation and transport devices.

Recessive: Refers to the member of a pair of alleles that fails to be expressed in the phenotype of the organism when the dominant member is present. Also refers to the phenotype of an individual that has only the recessive allele.

Recessive: Refers to the member of a pair of alleles that fails to be expressed in the phenotype of the organism when the dominant member is present. Also refers to the phenotype of an individual that has only the recessive allele.

Recombinant DNA Technology: A body of techniques for cutting apart and splicing together different pieces of DNA. When segments of foreign DNA are transferred into another cell or organism, the substance for which they code may be produced along with substances coded for by the native genetic material of the cell or organism. Thus, these cells become "factories" for the production of the protein coded for by the inserted DNA.

Recombinant DNA: Any DNA molecule formed by joining DNA segments from different sources. Recombinant DNAs are widely used in the cloning of genes, in the genetic modification of organisms, and in molecular biology generally.

Recombination: Process in which chromosomes or DNA molecules are broken and the fragments are rejoined in new combinations. Can occur in the living cell—for example, through crossing-over during meiosis—or in the test tube using purified DNA and enzymes that break and ligate DNA strands.

Reconnaisance Satellite: Also called a spy satellite because it is used to spy on other countries. It can provide intelligence information on military activities, detect missile launches or nuclear explosions, and pick up and record radio and radar transmissions while passing over a country. It can also be used as an orbital weapon by placing warheads on a low orbit satellite to be launched at a ground target, but this is not a recommended or frequent use of satellites.

Reconstruction: A crystal consists of a regular array of atoms, and the simplest model of a crystal surface would be generated by simply discarding all atoms to one side of a surface without changing the positions of the rest. In reality, however, the positions of the remaining atoms do change. A pattern of displacements that lowers the symmetry of the surface (relative to the ideally terminated crystal) is termed a surface reconstruction; some reconstructions alter the pattern of bonds.

Recrystallization: A process where materials crystallize repeatedly due to an annealing heat treatment. Recrystallization is the most common method employed for the purification of an impure solid. It is a process that relies on the use of the relative solubilities of the desired compound, and the impurities present, in the solvent chosen for use. The formation of a new set of strain-free grains within a previously cold-worked material due to an annealing heat treatment.

Rectifier: A device for converting AC into DC. Semiconductor rectifiers are diodes made large enough to disperse the heat caused by current flow. Device that converts bi-directional to one-way current flow.

Red Goo: Deliberately designed and released destructive nanotechnology, as opposed to accidentally created grey goo. [AS]

Redistribution: The modification of dopant concentration due to silicon oxidation. It occurs during diffusion.

Redox Group: A part of a molecule that undergoes changes in its electronic properties.

Reduced Mass: Many dynamical properties of a system consisting of two interacting masses, m_1 and m_2, are equivalent to those of a system in which one mass is fixed in space and the other has a mass (the reduced mass) with the value $m_1m_2/(m_1 + m_2)$. The reduced mass description has fewer dynamical variables.

Reduction (verb reduce): Addition of electron density to an atom, as occurs during the addition of hydrogen to a molecule or the removal of oxygen from it. Opposite of oxidation.

Reduction: The addition of one or more electrons to an atom, ion, or molecule.

Redunancy: The use of more components than are needed to perform a function; this can enable a system to operate properly despite failed components.

Re-entrant (behavior): A condition where a material retreats from its superconductive state and then re-enters it. This can be caused by a strong external magnetic field that dynamically exceeds the Hc of the material and/or is mis-aligned (in the case of some organic superconductors), a discordant temperature below Tc (in the case of some borocarbides), or by Jc hysteresis (momentarily exceeding the critical current density, causing the Tc to shift downward).

Reflection: (1) Deflection of a light beam at the interface between 2 media. (2) Bending of a light beam when passing from one medium to another, at different velocities of light.

Reflective Notching: Reflective Notching is a critical problem in making ICs in which the wafer typography reflects light into unexposed areas. This light may cause notching if it concentrates light in dark features that results in linewidth or transistor gate length contraction. It can be avoided by using anti-reflective coatings.

Reflow: A high-temperature process that makes the corners of an oxide insulating layer flow to other areas to improve step coverage.

Refraction: Bending of a light beam when passing from one medium to another, at different velocities of light.

Refractive Index: The ratio of the velocity of light in a vacuum to the velocity of light in a specific material. Using 1.0 as the base reference, the higher the umber, the slower the speed of the lightwaves in the material.

Refractory Metals: Metals that can withstand very high temperatures without melting. Tungsten, tantalum, titanium, and molybdenum are the refractory metals.

Refractory: A material that is very resistant to high temperature processes. The refractory metals in the periodic table include titanium, tungsten, tantalum, and molybdenum. The first three metals or their compounds are used in semiconductor processing extensively. A metal or ceramic that does not deteriorate rapidly or does not melt when exposed to extremely high temperatures.

Regenerative Medicine: A field of medicine concerned with developing and using strategies aimed at repair or replacement of damaged, diseased, or metabolically deficient organs, tissues, and cells via TISSUE ENGINEERING; CELL TRANSPLANTATION; and ARTIFICIAL ORGANS and BIOARTIFICIAL ORGANS and tissues. MeSH 2004

Register: A temporary storage location for an array of bits within a digital logic system.

Registration Overlay: The accuracy of the position of all patterns with respect to previous patterns that form other layers of a semiconductor wafer.

Relative Magnetic Permeability (μ_r): The ratio of the magnetic permeability of some medium to that of a vacuum (unitless), or: $\mu_r = \mu/\mu_0$, where μ_0 is the permeability of vacuum, a universal constant, which has a value of 1.257×10^{-6} H/m.

Relaxation Time: A measure of the rate at which a disequilibrium distribution decays toward an *equilibrium* distribution. The electron relaxation time in a metal, for example, describes the time required for a disequilibrium distribution of electron momenta (e.g., in a flowing current) to decay toward equilibrium in the absence of an ongoing driving force and can be interpreted as the mean time between scattering events for a given electron.

Release Etch: The process step in which all of the sacrificial material is removed to free the structural layers and achieve movable microstructures.

Released Layers: The patterned layers containing free-standing mechanical structures left behind by the removal of the sacrificial layer in surface micromachining.

Reliability (life, multi-use vs. single, calibration vs. accuracy drift): How well a sensor maintains both precision and accuracy over its expected lifetime. Also includes the robustness of the sensor.

Reliability Analysis: The probability of losing a required function as a result of component failures. Reliability analysis typically uses physical component models characterized by failure-rate statistics. The goal of the analysis is to calculate system-level failure rates for selected functionalities.

Reliability Base Events: The lowest-level faults on the reliability fault tree are considered base events.

Reliability Fault Tree: The fault tree represents the logical relationship between a top event and fault modes in the form of an AND-OR tree, based on the probability of losing a required function as a result of component failures.

Remote Sensing Satellite: A type of satellite that performs remote sensing from space. Remote sensing satellites generally monitor important resources for humans. For example, they might be used to track animal migration, locate mineral deposits, watch agricultural crops for weather damage, or see how fast the forests are being cut down. Because they are in space, remote sensing satellites are ideal for monitoring areas with harsh climates or difficult terrains.

Remote Sensing: The process of collecting data about something from a point far away. A satellite making observations of the Earth, therefore, is using its equipment for remote sensing.

Repeatability: The ability of an instrument to give, under specific conditions of use, closely similar responses for repeated applications of the same stimulus. The exactness with which a measuring instrument repeats indications when it measures the same property under the same conditions.

Repeater: An instrument that receives and retransmits signals anytime the satellite is close enough to be in contact with a ground station.

Replication: Duplication of an original using casting, embossing, or molding. The mechanism used in living systems to make copies of genetic information.

Replicator: A system able to build copies of itself when provided with raw materials and energy. Any system that can build copies of itself when provided with the appropriate raw materials and energy. In discussions of evolution, a replicator is an entity (such as a gene, a meme, or the contents of a computer memory disk) which can get itself copied, including any changes it may have undergone. In a broader sense, a replicator is a system which can make a copy of itself, not necessarily copying any changes it may have undergone. A rabbit's genes are replicators in the first sense (a change in a gene can be inherited); the rabbit itself is a replicator only in the second sense (a notch made in its ear can't be inherited). In discussions of evolution, a replicator is an entity (such as a gene, a meme, or the contents of a computer memory disk) which can get itself copied, including any changes it may have undergone. In a broader sense, a replicator is a system which can make a copy of itself, not necessarily copying any changes it may have undergone. A rabbit's genes are replicators in the first sense (a change in a gene can be inherited); the rabbit itself is a replicator only in the second sense (a notch made in its ear can't be inherited).

Reporter gene: Reporter Gene detection has become invaluable in the study of gene function and related cellular events. In these studies, the reporter gene acts as a surrogate for the coding region of the gene under study. Upon transfection of the reporter construct, expression of the reporter gene may be monitored by direct (e.g. enzymatic) or indirect (e.g. immunochemical) assay of the reporter protein. Localized expression of the reporter protein can be detected via immunochemistry using antibodies specific for the expressed reporter protein.

Representative Point: The point in a configuration space that represents the geometry of a system.

Reproducibility (of measurements): The closeness of the agreement between the results of measurements of the same measurand, where the individual measurements are carried out, changing conditions like (a) the method of measurement; (b) the observer; (c) the measuring instrument; (d) the location; (e) conditions of use; (f) time.

Reproducibility: A property of an experiment or process where one tends to receive consistent results from following a specific procedure.

Requirement: A translation of the needs into a set of individual quantified or descriptive specifications for the characteristics of an entity to enable its realization and examination.

Residual Stress: A persisting stress in a material free of external forces or temperature gradients.

Residual Stress: Stresses in a material that is free of external forces or temperature gradients and depends on the interactions between the defects within the component and the stresses to which it is exposed. These stresses are a combination of those applied in service and those that developed during manufacture and processing.

Resin: A semi-solid transparent flammable natural organic substance used in varnishes, printing inks, plastics etc.

Resist: A material or coating that can protect a surface from chemical reactions.

Resistance [Ù-ohm]: Characteristic of a resistor: in a 1-ohm resistance a current of 1 A produces a voltage drop of 1 V.

Resistance: The opposition of a material to the flow of electrical current through it. Energy lost due to resistance is a result of vibrations at the molecular level and manifests itself as heat in proportion to the square of the current flow. In a superconductor all resistance disappears below a certain temperature. However, this applies only to direct current (DC) electricity. Other types of losses result when transporting alternating current (AC). Examples of this include hysteresis, reactive-coupling and radiational losses. In the new high temperature ceramic superconductors, the power loss in applications like transmission lines is inversely proportional to the critical current density for low magnetic field applications. This limitation can be compensated for to some degree by increasing the ratio of voltage to current. In Type 2 superconductors carrying high-frequency alternating current, "skin effect" losses also result as the energy tends to migrate to the surface where the conductive medium is incontiguous, producing a pseudo-resistance. In some materials the amount of resistance may also depend on the direction of current flow (anisotropic resistivity) and/or presence of an external magnetic field (hall effect). The property of a body whereby it opposes and limits the passage through it of a steady electric current. Its units are ohms.

Resistive Loading: The reduction in the voltage at the point being probed. This is caused when a probe is introduced into a circuit. Some of the current that was flowing in the circuit will now flow into the probe.

Resistor: Energy dissipative element consisting of a poor conductor in series with connecting wires. Devices designed to have definite amounts of electrical resistance and are used in circuits to limit current flow or to provide voltage drops.

Resolution (of an indicating device): A quantitative expression of the ability of an indicating device to distinguish meaningfully between closely adjacent devices.

Resolution: (1) The minimum distance between two objects that can be distinguished in microscopy or the minimum spacing between two features that can be fabricated with lithography. The smallest measurable change in input that will produce a small but noticeable change in the output. In the context of chemical separations, defines the completeness of separation. (2) The smallest measurable change in input that will produce a small but noticeable change in the output. In the context of chemical separations, defines the completeness of separation.

Resonant Circuit: A circuit that contains both inductance and capacitance elements to achieve resonance at a frequency that can be manipulated by altering the value of the inductance or capacitance.

Resonant Frequency: In an electrical circuit, the frequency that exists when the inductive reactance and the capacitive reactance are of equal magnitude, causing electrical energy to oscillate between the magnetic field of the inductor and the electric field of the capacitor. At this frequency, the circuit has a maximum output for a given input. Resonance frequency. The frequency at which a moving member or a circuit has a maximum output for a given input.

Resonant Tunneling Devices (RTDs): Here we talk about the future nanoelectronic devices using quantum effects employed in semiconductor devices for fabrication of computer devices and high frequency oscillators. The underlying principles used for defining the perpendicular transport of electrons in low dimensional structures up to room temperature and its current and voltage characteristics are explained using energy band diagrams. MOSFETS with its special characteristic features serves two purposes. First case its acts as an excellent conductor device even with its scaling properties and even with small changes in its bias voltage results in large changes in its conductivity, hence its acts as an amplifier device. Though it functions as a two state device there some factors need to be considered for its implementation. The obstacles associated with MOSFET are high electric field, heat power dissipation, vanishing bulk properties, shrinkage of depletion layers and unevenness of the depletion layer. Corrections have to be applied for these devices. Once the devices approach the nanoelectric properties, the bulk properties of semiconductor devices can be replaced by quantum physics which depend on the electron properties, energy quantization effects and tunneling come into effect. When quantum effects are applied to semiconductor devices the consideration of doped materials no longer considered as major issue. These devices are of great importance for high speed electronics. Here we discuss the quantum properties, general used terms such a quantum dot, tunneling effect, island, subband and electron behavior in quantized energy levels to deal with the resonant tunneling devices and the application of the quantum properties in various devices such as Resonant Tunneling Diodes (RTDs), Resonant Tunneling Transistors (RTTs). Here we explain the theory related to tunneling and the mathematical derivation for transmission probability of the particle in the double barrier junction.

Focus primarily on explaining the operation of resonant tunneling devices, because they employ quantum effects in their simplest form. Presently, these devices usually are fabricated from layers of two different III/V semiconductor alloys, such as the pair GaAs and AlAs. The simplest type of resonant tunneling device is the resonant tunneling diode (RTD). A resonant-tunneling diode is made by placing two insulating barriers in a semiconductor, creating between them an island or potential well where electrons can reside. Resonant tunneling diodes are made with center islands approximately 10 nanometers in width. Whenever electrons are confined between two such closely spaced barriers, quantum mechanics restricts their energies to one of a finite number of discrete "quantized" levels. This energy quantization is the basis for the operation of the resonant-tunneling diode. The only way for electrons to pass through the device is to "tunnel," quantum mechanically, through the two barriers. The probability that the electrons can tunnel is dependent on the energy of the incoming electrons compared to the energy levels on the island of the device. If the energy of the incoming electrons differs from the energy levels allowed inside the potential well on the island, then current does not flow. The device is switched "off."

By adding a small gate electrode over the island of an RTD one may construct a somewhat more complex resonant tunneling device called a resonant tunneling transistor (RTT). In this three-terminal configuration, a small gate voltage can control a large current across the device. Because a very small voltage to the gate can result in a relatively large current and voltage across the device, amplification or "gain" is achieved. Thus, an RTT can perform as both switch and amplifier, just like the conventional bulk effect transistor. Unlike conventional bulk effect transistors which usually have only two switching states, "on" and "off," resonant tunneling devices like RTDs and RTTs can have several switching states. This occurs because the quantum well on the island may exhibit several possible energy levels. Thus, as the voltage bias on an RTD is increased from zero, where it is "off," the device initially switches "on" when the first energy level comes in resonance with the incident electrons. Then, it switches off again as the bias is increased further, past resonance. However, if there is a second energy level in the quantum well, then the device can switch on again as the bias voltage of the RTD (or the gate voltage of an RTT) is increased still further. This multistate switching behavior permits each device to "count higher" and represent more logic states.

Resonator: A mechanical structure that vibrates, sometimes at a resonance frequency.

Respirocrit: In medical nanorobotics, the volume-fraction or bloodstream concentration of respirocyte nanorobots, expressed as a percentage.

Respirocyte: In medical nanorobotics, a theorized bloodborne spherical 1-micron (nanorobotic) device having a 1000-atm pressure vessel with active pumping powered by endogenous serum glucose, that serves as a mechanical artificial red blood cell.

Response Characteristic: The relationship between a stimulus and the corresponding response.

Response Time: The time interval between the instant when a stimulus is subjected to a specified abrupt change and the instant when the response reaches and remains within specified limits of its final steady value.

Restriction Enzyme: An enzyme that cuts DNA at a specific site, allowing biologists to insert or delete genetic material.

Result: If the mass difference between and is not very small (more than 0.1 eV), there is no more than 10% of mixing between the neutrinos, that is has not more than 10% of chance to transform into a . But the history does not stop there, because the neutrino is facetious. Since 1969, a physicist named Ray Davis tried, in Homestake mine, under 3000 meters of earth and stones, to catch a few solar neutrinos per year, using a very big detector made of 600 tons of industrial solvent based on chlorine. His first results surprised. They are confirmed today, after 20 years of data: above a neutrino energy of 1 MeV, the sun emits three times less neutrinos than predicted by the standard solar model. Astrophysicists scraped their heads and other experiments were undertaken in order to confirm this unexpected deficit. Especially, three experiments: GALLEX, SAGE and KAMIOKANDE. The experiment of Davis, HOMESTAKE, uses chloride, GALLEX uses gallium and KAMIOKANDE uses water. Theoretically, the different experiment does not see the same neutrinos, according to their origin from thermonuclear reactions in the sun. Could the observed deficit of solar neutrinos come from neutrinos oscillation?... The idea is attractive, but the results of the experiments show that it is difficult to make it practical. On an other hand, in 1985, S.P. Mikheyev and A.Y. Smirnov developped the work of L. Wolfenstein about neutrino oscillation enhanced by the presence of matter: this is the MSW effect. The solar neutrino deficit could come also from an oscillation during the path inside the sun. But experiment is the only truth. Neutrinos are re-CERNed (1989-1996...) In 1989, as early as the first months of data taking at LEP, the new CERN particle accelerator, the study of the Z boson lifetime allows to show that only three light neutrino families exist. This was a major physics result! The shorter is the lifetime of a particle, the more undetermined is its mass. One says that its mass distribution has a width. And this width increases with the number of possibilities for the particle to decay. The Z boson, whose mean lifetime is around 10^{-23} second, can decay in pairs particle, anti-particle like neutrino, anti-neutrino. The more is the number of neutrino families, the larger is the width of the Z mass distribution. In 1992, two experiments at CERN are built in order to detect neutrino oscillations: NOMAD and CHORUS, which, thanks to neutrino oscillations, hope to see some neutrinos inside a beam of produced with the protons of the CERN SPS accelerator. The data taking began in 1994 and first results were given in 1998. The neutrino still dance (1993-1998...) In 1995, the GALLEX experiment publishes its final result: a deficit of about 40% over almost the whole spectrum of solar neutrinos. SAGE confirms wisely this number and KAMIOKANDE observes a deficit of 50%, but for the Bore neutrinos only (above 7.5 MeV). The HOMESTAKE experiment, sensitive to Bore and Beryllium neutrinos had seen a deficit of 70%. The mystery is still there. Depending on the solar neutrino energy, the deficit is not the same. The MSW effect could explain this selective deficit. Finally, KAMIOKANDE observes a deficit of in the cosmic rays showers. To try to interpret all those deficits by using only neutrino oscillations interpretation is not an easy work!... And it becomes very difficult if you try to incorporate the results obtained by LSND experiment at Los Alamos: an oscillation between anti- and anti-. In 1996, LSND experiment announced new results: 22 interactions of anti- when 4 such interactions were expected. The non zero neutrino mass hypothesis seems to be stronger, but the scientific quest is like a mountain climber: it must make sure of its hanging points before continuing to progress. Results of other experiments and confirmation from LSND are thus impatiently waited. In 1998, the LSND confirmation arrives and a Japanese experiment, Super-

Kamiokande, has seen also an anomaly in the number of atmospheric neutrinos and solar neutrinos. The results are still preliminary but all seems in place for a massive neutrino or for a new gift from nature (that is a new scientific discovery).

Return on Assets: Net income divided by net assets. This measure of profitability is considered to be broader than a return on equity because it takes into account both equity and debt financing.

Return on Investment: At the firm level, it is often defined as net income divided by a shareholder's equity.

Reverse Bias: If an external voltage is applied across a p-n junction with a polarity that p-type material is negative with respect to the n-type material, the junction is under reverse bias. The insulating bias for a p-n junction rectifier; electrons flow into the p side of the junction.

Reverse Engineering: The process of determining the actual circuit by study of the geometrical layout of a fabricated chip.

Reverse Pharmacology: [Masashi] Yanagisawa [Howard Hughes Medical Institute at the University of Texas Southwestern Medical Center at Dallas] went after these receptors because they are mostly "orphan receptors"— those with no known ligand. He suspected that the ligand for many of these receptors would turn out to be a peptide hormone. The computer research yielded about 50 sequences that the group felt were likely to be G protein- coupled receptors, and then set about using those receptors as bait to capture peptide hormones, their true quarry. This strategy is known in the field as "reverse pharmacology. "In traditional pharmacological research, the hormone is identified first," Yanagisawa said. "That hormone is then used as a tag to pull out the receptor molecule. We're doing this in reverse."

RF (Radio Frequency): Alternating voltages and currents having wavelengths between 22 km (lowest energy) and 1mm (highest energy). The AC current is input to an antenna and an electromagnetic (EM) field is generated suitable for wireless broadcasting.

RF Sputtering: A sputtering system with a cathode and anode configuration and an RF source that can sputter metal or insulator thin films in inert or reactive atmospheres in which DC methods cannot.

RF: Radio Frequency — refers to alternating voltages and currents having frequencies between 9 kHz and 3 MHz.

RHEED: Reflection High Energy Electron Diffraction.

Rheology: The study of the ability of matter to flow or deform. Studies for the successful operation of sensors.

Ribonuclease: An enzyme that cuts RNA molecules into smaller pieces.

Ribonuclease: An enzyme that cuts RNA molecules into smaller pieces.

Ribosome: A molecular machine, found in all cells, which builds protein molecules according to instructions read from RNA molecules. Ribosomes are complex structures built of protein and RNA molecules. A naturally occurring molecular machine that manufactures proteins according to instructions derived from the cell's genes.

RIE (Reactive Ion Etching): Dry etching by plasma having chemically active gas ions.

Rigid structure: As used in this volume, a *covalent* structure that is reasonably stiff. In a typical rigid structure, all modes of deformation encounter first-order restoring forces resulting from some combination of bond stretching and angle bending; such a structure cannot undergo deformation by bond torsion alone. Meeting this condition usually requires a *polycyclic diamondoid* structure.

Rinse: The use of ultra-pure water to neutralize and remove pollutants from a wafer.

Riser: A cavity that contains molten metal provided to compensate for the contraction of the metal in a casting as it solidifies.

RMS (Root Mean-Square): The square root of the arithmetic average of a set of squared instantaneous values. It refers to the most common mathematical method of defining the effective voltage or current of an AC wave.

RNA: Ribonucleic acid; a molecule similar to DNA. In cells, the information in DNA is transcribed to RNA, which in turn is "read" to direct protein construction. Some viruses use RNA as their genetic material.

RNAi RNA Interference: A gene silencing phenomenon whereby specific dsRNAs (RNA, DOUBLE-STRANDED) trigger the degradation of homologous mRNA (RNA, MESSENGER). The specific dsRNAs are processed into SMALL INTERFERING RNA (siRNA) which serves as a guide for cleavage of the homologous mRNA in the RNA- INDUCED SILENCING COMPLEX (RISC).

Roadmap: The semiconductor industry publishes a estimate of when new technologies will become available for use by the microelectronics industry.

Robot: The word 'robot' was coined by the Czech playwright Karel Capek (pronounced "chop'ek") from the Czech word for forced labor or serf. Capek was reportedly several times a candidate for the Nobel prize for his works and very influential and prolific as a writer and playwright. Mercifully, he died before the Gestapo got to him for his anti- Nazi sympathies in 1938. The use of the word Robot was introduced into his play *R.U.R.* (Rossum's Universal Robots) which opened in Prague in January 1921. The play was an enormous success and productions soon opened throughout Europe and the US. R.U.R's theme, in part, was the dehumanization of man in a technological civilization. You may find it surprising that the robots were not mechanical in nature but were created through chemical means. [Comp-AI Robotics FAQ]

Robotic System: Automated device where materials are transferred by the physical movement of a delivery device relative to the ultimate receptacle, or vice versa.

Robotics: "A reprogrammable, multifunctional manipulator designed to move material, parts, tools, or specialized devices through various programmed motions for the performance of a variety of tasks" Robot Institute of America, 1979 Obviously, this was a committee-written definition. It's rather dry and uninspiring. Better ones for 'robotics' might include: Force through intelligence. Where AI meet the real world. ...refers to the study and use of robots.

Robust Pipelines: Best identified in retrospect. Predicting which drugs will be most profitable is difficult (if not impossible). Drug pipelines can be filled with promising products, but successful use in humans is the ultimate validation.

Robust: A process which is relatively insensitive to human foibles and variables in the way (for example, an **assay**) is carried out, a statistical term.

Rod Logic: One study not conducted by Ames or partners is particularly worth mentioning since it places a loose lower bound on the computational capabilities of molecular nanotechnology. Drexler designed a number of computer components using small diamondoid rods with knobs that allow or prevent movement to accomplish computation. While this tiny mechanical Babbage Machine is probably not an optimal computational engine, its calculated performance for a desktop computer is 10^{18} MIPS — about a million times more powerful than the largest supercomputer that exists today. Note that with very fast computation energy use and heat dissipation become a severe problem. One approach to addressing this issue is reversible logic.

ROM (Read-Only Memory): A device that permanently stores information repeatedly used-such as tables of data, characters for electronic displays, etc. Unlike a RAM, a ROM cannot be altered.

Room-temperature Superconductor: There are NO confirmed room-temperature superconductors (as was once reported for lithium-beryllium-hydride and for lead-silver-carbonate). However, it has been theorized that a metallic form of hydrogen might be a room-temperature superconductor. In 1996 physicists at Lawrence Livermore Laboratory were able to briefly create metallic hydrogen. But, its existence was fleeting and no measurement of the Meissner effect was possible. Zero resistance *has* been observed at room temperatures in ballistic quantum wire. However, having one-dimensional geometry, this wire does not exhibit the Meissner effect, except when configured as a closed loop.

Rotaxanes: Rotaxane family of molecular machines is characterized by two parts —a dumb-bell shaped compound with two heavy chemical groups at the ends and a light, cyclic component, called macrocycle, interlocked between the heads. It has been shown that a reversible switch can be made with a rotaxane setup. For this, one needs to have two chemically active recognition sites in the neck region of the dumb-bell. In this particular example, the thread was made of polyether, marked by recognition sites hydroquinol units and terminated at the ends by large triisoproplylsilyl groups. A tetracationic bead was designed and self-assembled into the system that interacts with the recognition sites. The macrocycle has a natural low energy state on the first recognition site, but can be switched among the two sites reversibly upon application of suitable stimuli. Depending on the type of rotaxane setup the stimuli can be chemical, electrochemical or photochemical. The stereo-electronic properties of the recognition sites can be altered protonation or deprotonation, or by oxidation or reduction, thereby changing their affinity towards the macrocycle. In a recent example, light induced acceleration of rotaxane motion was achieved by photoisomerization, while similar controls through alternating current (oscillating electric fields) was shown before.

Roughing Pumps: Pumps used to achieve a vacuum of not greater than 10^{-8} torr. They can be used to evacuate chambers in a cluster tool.

Router: A device or software that determines the next network point to which data should be forwarded on the way to its final destination.

Run Card: A form (paper or electronic) summarizing some work that needs to be done to move a run forward. Like a process sequence, a run card might describe the whole run from start to finish, or some subset of it. Common subsets include the work to be done at just one fab, or the work to be done at one fab plus the work done at all previous fabs involved in that run.

Run: A process sequence together with a description of the tangibles to be used in processing (masks, wafers) and all the logistical minutiae required to make it actually happen—process approval, scheduling, tracking, billing, etc.

S

S phase: Period of a eucaryotic cell cycle in which DNA is synthesized.

Saccharomyces: Genus of yeasts that reproduce asexually by budding or sexually by conjugation. Economically important in brewing and baking, they are also widely used in genetic engineering and as simple model organisms in the study of eucaryotic cell biology.

Sacrificial Anode: An active metal or alloy that corrodes and protects another metal or alloy to which it is electrically coupled.

Sacrificial Layer: A thin film that is later removed to release a microstructure from its substrate.

Sacrificial-Layer Etching: Removal of a buried fast-etching layer used, e.g. to create freely movable structures.

Salmonella: Rod-shaped, motile, aerobic genus of bacteria. Includes species that cause food poisoning.

Salt Bridge: An ionic bond between charged groups that are part of larger covalent structures; salt bridges occur in many proteins.

SAMFET: (self assembled monolayer field effect transistor).

Where a few molecules act as FETs, exhibiting both very strong gain, and extraordinarily rapid response.

Sample Prep: Sample preparation isn't the sexiest job around, but it is one of the most critical. The quality of isolated nucleic acid and protein samples is critical to generating accurate and informative data. As genomic and proteomic technologies move in the direction of higher throughput, upstream sample preparation becomes a potential bottleneck. Sample capture, transportation, storage, and handling are as critical as extraction and purification procedures. Obtaining homogenous samples or isolating individual cells from clinical material is imperative. Standards are essential. Advances in microfluidic and microarray technologies have further amplified the need for higher throughput, miniaturized, and automated sample preparation processes.

Sample: One or more items taken from a population or a process and intended to provide information on the population or process. Portion of material selected from a larger quantity in some manner chosen so that the portion is representative of the whole. In statistics, a group of individuals often taken at random from a population for research purposes.

Sampling Interval: The size of the samples used to measure something, e.g. in imaging sampling, it refers to pixel size. In spectroscopy, sampling refers to the smallest spectral bandwidth used to measure something. Sampling, as applied to an analog-to-digital converter, is the process that transforms a continuous function into a series of discreet values at a linear time rate.

Sanguinatation: In medical nanorobotics, locomotion (especially swimming by a nanorobot) through the bloodstream.

Sarcomere: Repeating unit of a myofibril in a muscle cell, composed of an array of overlapping thick (myosin) and thin (actin) filaments between two adjacent Z discs.

Sarcoplasmic Reticulum: Network of internal membranes in the cytoplasm of a muscle cell that contains high concentrations of sequestered Ca2+ : that is released into the cytosol during muscle excitation.

Satellite DNA: Regions of highly repetitive DNA from a eucaryotic chromosome, usually identifiable by its unusual nucleotide composition. Satellite DNA is not transcribed and has no known function.

Satellite: Something that is in orbit around something else. The Moon is a satellite of the Earth, for example. An *artificial* satellite is a satellite that was constructed and placed in orbit by humans.

* scientific notation: The system used when talking about very small or very large quantities. In scientific notation there is always a base unit, for example metre, Hertz, gram. Instead of using lots of zeros, scientific notation uses prefixes to indicate multiples of ten. Below is a chart of these prefixes, what they mean, and what they might measure: 10^9 giga—the number of Hertz at which satellites transmit signals
* 10^6 mega—the distance around the Earth
* 10^3 kilo—a person's weight
* 10^{-2} centi—the width of a computer monitor
* 10^{-3} milli—the head of a pin

* 10^{-6} micro—the width of human tissue
* 10^{-9} nano—the wavelength of visible light
* 10^{-12} pico—the width of a virus

Saturated Molecule: Molecule containing carbon-carbon bonds that has only single covalent bonds.

Saturated: An organic molecule is described as saturated if it is a closed shell species lacking *double* or *triple bonds*; forming a new bond to a saturated molecule requires the cleavage of an existing bond.

SC1 (Standard Clean 1): The first cleaning step of the RCA clean. SC1 is done in a megasonic bath containing ammonium-hydroxide and hydrogen-peroxide. This step removes organic residues and particles.

SC2 (Standard Clean 2): The second cleaning step in the RCA clean. SC2 is a heated mixture of hydrochloric-acid and hydrogen-peroxide and removes metal contaminants.

Scalable: Capable of being expanded for high- throughput. Analogous to recipes optimized for large groups, rather than standard recipes being quadrupled or more, with less than ideal results. Also spelled scaleable.

Scaling: A technique of reducing the size of an existing IC die by selective shrinking of the X and/or Y dimensions.

Scanning Capacitance Microscopy (SCM): A variation of scanning force microscopy that maps the localized capacitance of a surface. A method for mapping the local capacitance of a surface. A microscope producing an image by using reflected electron beams that scan the surface of a specimen. A method of producing images of a surface by scanning an electron beam over the sample and measuring the electronic interactions with the interface. One or more items taken from a population or a process and intended to provide information on the population or process. Portion of material selected from a larger quantity in some manner chosen so that the portion is representative of the whole.

Scanning Electron Microscopy (SEM): Any analytical technique which involves the generation and evaluation of secondary electrons (and to a lesser extent back scattered electrons) by a finely focused electron beam (typically 10 nm or less) for high resolution and high depth of field imaging. Microscopy in which the object is examined directly by an electron beam scanning the specimen point-by- point, giving the surface image a three-dimensional quality. MeSH, 1972

Scanning Force Microscope (SFM): (1) An instrument able to image surfaces to molecular accuracy by mechanically probing their surface contours. A kind of proximal probe. .... A device in which the deflection of a sharp stylus mounted on a soft spring is monitored as the stylus is moved across a surface. If the deflection is kept constant by moving the surface up and down by measured increments, the result (under favorable conditions) is an atomic-resolution topographic map of the surface. Also termed an atomic force microscope. (2) Since its invention in 1986 by Binnig et al., scanning force microscopy has become a powerful tool for the investigation of surfaces. A great advantage of this method is the high resolution—down to the atomic scale—which can be achieved even in air and even on insulating surfaces. However, the greatest advantage is the possibility to investigate surfaces *in situ* in a liquid phase.

Scanning Near-field Optical Microscopy (SNOM) : A type of scanning probe microscopy that maps the near-field optical properties of an interface.

Scanning Probe Microscopy (SPM): (1) A method for imaging nanoscale features of surfaces by scanning a sensor (probe) over a surface. Near-field effects such as tunneling, van der Waals forces, local fields and more are serially detected at localized points on the surface and used to create an SPM image. (2) Electron microscopy in which a very sharp probe is employed in close proximity to a surface, exploiting a particular surface- related property. When this property is local topography, the method is atomic force microscopy, and when it is local conductivity, the method is scanning tunneling microscopy.

Scanning Probe Microscopy: Electron microscopy in which a very sharp probe is employed in close proximity to a surface, exploiting a particular surface-related property. When this property is local topography, the method is atomic force microscopy, and when it is local conductivity, the method is scanning tunneling microscopy.

Scanning Projection Aligner: An aligner that scans light through a slit across a mask to produce an image on a wafer. Since scanning is done through a slit, a relatively small lens can be used. The technique allows larger areas to be imaged without requiring huge optical systems.

Scanning Thermal Microscopy (SThM): A type of scanning probe microscopy that maps the local temperature and thermal conductivity of an interface.

A method for observing local temperatures and temperature gradients on a surface.

Scanning Transmission Electron Microscopy (STEM): A special TEM- technique in which an electron transparent sample is bombarded with a finely focused electron beam (typically of a diameter of less than 10 nm) which can be scanned across the specimen or rocked across the optical axis and transmitted secondary, backs scattered and diffracted electrons as well as the characteristic X-ray spectrum can be observed. STEM essentially provides high resolution imaging of the inner microstructure and the surface of a thin sample (or small particles), as well as the possibility of chemical and structural characterization of micrometer and sub- micrometer domains through evaluation of the X-ray spectra and the electron diffraction pattern.

Scanning Tunneling Microscope (STM): A device designed to image conducting surfaces of materials with atomic precision. Invented by Binning and Rohrer in 1981 which were awarded the Nobel Prize in Physics in 1986 for this discovery. The basic element of the STM is a sharp tip made of wolfram, platinum, iridium, or some other conducting material. The tip is typically a specially prepared wire that has been sharpened by special chemical and/or mechanical procedure, so that, ideally, only one or few atoms are located at the tip end. Carbon nanotubes are also used for STM tips. The tip is controlled by a computer which enables its precise positioning with respect to the surface which is to be imaged. The tip has to be appropriately positioned within a manipulator which is connected to a computer. A bias voltage is applied between the tip and the material, which results in a small current when the tip is very close to the surface (nanometer and smaller), but only if the material investigated is conducting. The whole experimental setup should be in vacuum. The current changes as the tip is moved across the surface (*x-y* plane). In one mode of operation (constant height mode), the distance between the tip and the surface is kept constant (constant *z* coordinate of the tip) and the changes in the current are recorded. Variation of current as a function of the tip position (x,y) can be interpreted as the image of the surface. In another operation mode, the current is kept constant and the variation of *z* coordinate is recorded. The current varies in *x-y* plane due to the fact that the surface of material is inhomogeneous on the atomic scale, i.e. the electronic densities are not everywhere the same. In order for current to be established, the electrons from the surface must travel towards the tip or vice versa. Since the space between the tip and the surface is essentially vacuum, i.e. it is nonconducting, the electrons tunnel through this space. Such a behavior can be explained only by quantum theory. Since the tunneling current decreases approximately exponentially with the tip-surface separation, the tip has to be extremely close to the surface in order to record the current, which is typically given in nanoampere units.

Schottky Contact: A contact created when metal is directly in contact with a semiconductor. It does not allow current both ways and metal silicides like $TiSi_2$, to ensure that current flows freely both sides. The properties of the Schottky contactare determined by the properties of the majority carriers in the semiconductor and are used as rectifiers in the form of a Schottky diode.

Schottky Defect: The presence of a missing atom in a crystal lattice. Also known as vacancy.

Schottky Diode: Possesses a lower voltage drop than conventional diodes and is an order of magnitude faster than the conventional PN diodes and hence used in RF applications. A normal diode has between 0.7-1.7 volt drops, while a Schottky diode voltage drop is between approximately 0.5-0.45 volts, which accounts for higher system efficiency. Walter H. Schottky is a semiconductor diode with a low forward voltage drop. A typical application is discharge-protection for solar cells connected to lead-acid batteries. While standard silicon diodes have a forward voltage drop of about 0.6 volts, Schottky diodes have a drop of only about 0.2 volts. A Schottky diode uses a metal-semiconductor junction as a Schottky barrier (instead of a semiconductor-semiconductor junction as in conventional diodes). This Schottky barrier results in both very fast switching times and low forward voltage drop.

These have a very low forward voltage drop, usually 0.15 to 0.45 V, which makes them useful in battery-powered and low-voltage circuits. Also in mixer circuits for RF.

Schottky TTL: A form of TTL logic in which Schottky diodes are used to clamp the transistors out of saturation, effectively eliminating the storage of charge within the transistor—allowing increased switching speeds.

Schwann Cell: Glial cell responsible for forming myelin sheaths in the peripheral nervous system.

Science Court: A name (originally applied by the media) for a government-conducted fact forum.

Science: The process of developing a systematized knowledge of the world through the variation and testing of hypotheses. The pursuit of knowledge and understanding, from the Latin term *scientia*, which means 'knowledge'.

Scintillation: Burst of luminescence of short duration caused by an individual energetic particle.

SCREAM: Single Crystal Reactive Ion Etching and Metallization.

Scribe Line: The partition between neighboring die on the wafer. It is used as the cutting area in sawing a wafer into the individual die. Also referred to as saw lane or street.

Sealed Assembler Lab: A general-purpose assembler system in a container permitting only energy and information to be exchanged with the environment.

Sealed Assembler Laboratory: A work space, containing assemblers, encapsulated in a way that allows information to flow in and out but does not allow the escape of assemblers or their products.

Search and Rescue Satellite: A type of satellite used to help locate a ship, airplane, or individual lost or in distress in a remote area. Search and rescue satellites use devices like radio signals coming from beacons, the doppler effect, and radar to calculate where a distress signal is coming from. Usually, search and rescue satellites operate in fleets of several satellites. That way, the satellites can cover a large area, and they can communicate with one another.

Second Messenger: Small molecule that is formed in or released into the cytosol in response to an extracellular signal and helps to relay the signal to the interior of the cell. Examples include cAMP, IP3 and Ca2+.

Secondary Electron Imaging: Production of secondary electrons is very topography related. Due to their low energy, 5eV, only secondaries that are very near the surface (<10nm, see Interaction Volume) can exit the sample and be examined. Any changes in topography in the sample that are larger than this sampling depth will change the yield of secondaries due to collection efficiencies. Collection of these electrons is aided by using a "collector" in conjunction with the secondary electron detector. The collector is a grid or mesh with a +100V potential applied to it which is placed in front of the detector, attracting the negatively charged secondary electrons to it which then pass through the grid-holes and into the detector to be counted. When a Secondary Electrons collide with the solid-state semiconductor detector an electron-hole pairs are created which are then counted. This quantity is translated into a pixel intensity and displayed on the CRT, forming the image. See the SEM page for details on image formation.

Secondary Electron: Produced by an incident electron passing "near" an atom in the specimen, near enough to impart some of its energy to a lower energy electron (usually in the K-shell). This causes a slight energy loss and path change in the incident electron and the ionization of the electron in the specimen atom. This ionized electron then leaves the atom with a very small kinetic energy (5eV) and is then termed a "secondary electron". Each incident electron can produce several secondary electrons.

Secondary Structure: Regular local folding pattern of a polymeric molecule; in proteins, a helices and b-pleated sheets.

Second-Level Packaging: The assembly of the packaged IC component into the next level, such as a rinted-circuit board (PCB). During second-level packaging, through-hole components like PGAs account for more than 50% of the IC packaging on a PCB.

Secretory Vesicle: Membrane-bounded organelle in which molecules destined for secretion are stored prior to release. Sometimes called secretory granule because darkly staining contents make the organelle visible as a small solid object.

Seeds Model: A probability model for converting defect density to predicted yield.

Selective Etching: The etching of one material without removing any other material.

Selectivity: (1) The ability of a sensor to measure only one metric or, in the case of a chemical sensor, to measure only a single chemical species. (2) The extent to which a compound hits the intended *drug target.*

Self Assembly: Self assembly creation of material from its constituent components in a spontaneous, 'natural' manner, i.e. by an interaction between the components or by a specific rearrangement of them, that proceeds naturally without any special external impetus. Also known as 'Brownian assembly'. A typical example are self-assembled monolayers where molecules (such as alkanethiol) spontaneously form an extremely well ordered layer only one molecule thick when deposited on substrates (e.g. Au (gold)). Another example is a self-assembly of proteins in protein structures (in cells) such as microtubules, actin filaments and similar. Although individual proteins stochastically move within a cell (Brownian motion), they eventually stick to a specific place as a properly positioned part of an ordered structure. This term is not related to self-assembler (or assembler) which is a speculative nano-fiction concept (nanobot). A specific type of assembler that makes use of self-assembly such that the assembly process would theoretically not require external energy or information input. In a self-assembler, the choice of starting materials (input) determines the assembly and is always of higher energy than the product (output).

Self-Assembled Monolayer (SAM): A two-dimensional film, one molecule thick, covalently assembled at an interface.

Self-Assembler: A specific type of assembler that makes use of self-assembly such that the assembly process would theoretically not require external energy or information input.

Self-assembling Biomolecular Materials: Examples of self- assembly include protein folding, the formation of liposomes, and the alignment of liquid crystals. While this type of equilibrium self- assembly is the central focus of this report, it is important to emphasize that much biological assembly is also driven by energy sources such as adenosine triphosphate (ATP), which power biomotors

Self-assembling Peptides: Using standard peptide chemistry, we are able to manufacture small oligopeptide fragments which self- assemble into nano- fibers on a scale identical to *in vivo* extracellular matrix and impossible to synthetically produce by other manufacturing techniques. This explains why most of the alternatives to our PuraMatrix are harvested from animal and human cadavers (bovine collagen, cartilage extract, intestinal submucosa, mouse tumor extract, bovine gelatin, cadaver tissue, etc.) with high processing costs, high quality variability, and disease transmission risks. Synthetic alternatives fail to match the extremely fine fibers, high water content, and neutral charge that is so important to cells growing, dividing, expressing proteins, and remodeling tissue both in and outside of the body.

Self-Assembly : A method of integration in which the components spontaneously assemble, typically by bouncing around in a solution or gas phase until a stable structure of minimum energy is reached.

A process in which supramolecular hierarchical organization is established without external intervention.... The approaches used can be expected to fall into two general categories. The first involves directly mimicking biological systems or processes to produce materials with enhanced properties. An example of this approach is the use of molecular genetic techniques to produce polymers with unprecedentedly uniform molecular length. The second category involves studying how nature accomplishes a task or creates a structure with unusual properties, and then applying similar techniques in a completely different context or using completely different materials. *Narrower terms:* self-assembling biomolecular materials, self-assembling peptides.

In chemical solutions, self-assembly (also called Brownian assembly) results from the random motion of molecules and the affinity of their binding sites for one another. Also refers to the joining of complementary surfaces in nanomolecular interaction. See MITRE Nanosystems Research Task: Self-Assembly of Nanosystems and Microsystems

Self-repair: Indicating ability to heal itself without outside intervention.

Self-replication: The act of objects making copies of themselves—similar to biological reproduction except that self-replicating objects should make exact copies of themselves.

More accurately labeled "exponential replication," self-replication refers to the process of growth or replication involving doubling within a given period. Example: create one assembler. Program it to create another, and program that one likewise, etcetera, until you have a speficied amount [which is the important part — how to make them STOP]. Self replication is an effective route to truly low cost manufacturing. Our intuitions about self replicating systems, learned from the biological systems that surround us, are likely to seriously mislead us about the properties and characteristics of artificial self replicating systems designed for manufacturing purposes. Artificial systems able to make a wide range of non- biological products (like diamond) under programmatic control are likely to be more brittle and less adaptable in their response to changes in their environment than biological systems. At the same time, they should be simpler and easier to design. The complexity of such systems need not be excessive by present engineering standards. The act of objects making copies of themselves—similar to biological reproduction except that self-replicating objects should make exact copies of themselves.

SEM (Scanning Electron Microscopy): Allows features, too small for optical microscopy to resolve, to be visible by scanning the sample with an electron beam.

Semiconductor: A substance or object with conductive properties between those of a conductor and a insulator.

Semiconductor Materials: Semiconductor materials are insulators at zero temperature that conduct electricity in a limited way at room temperature. The defining property of a semiconductor material is that it can be doped with impurities that alter its electronic properties in a controllable way. Because of their application in devices like transistors (and therefore computers) and lasers, the search for new semiconductor materials and the improvement of existing materials is an important field of study in materials science. The most commonly used semiconductor materials are crystalline inorganic solids. These materials can be classified according to the periodic table groups from which their constituent atoms come

Semiconductor Superlattices: When alternating layers of two semiconductors with different band gaps are deposited and thickness of these layers lies between 1-10 nm (which corresponds approximately to 10–100 atomic layers), what we obtain is semiconductor superlattice. The mostly known example of such a structure is GaAs/AlAs system depicted schematicallytogether with its corresponding potential profile. Differences in band gaps of materials lead to discontinuities in valence and conduction band and a set of square potential wells (low band gap semiconductor) separated by potential barriers (high band gap semiconductor) forms. Let us consider one of these wells from the quantum mechanical point of view. If we assume that an electron mean free path within a well is substantially longer than well width and the barrier width is infinite, what we obtain is a classical problem of a particle confined in a square well of finite depth. Without going to the details, it is sufficient to state that the discrete electron energy levels appear in the well and an increase, D, of the electron energy from the ground state to its first bound state can be calculated as the lowest order solution of:

$$\cos (kL/2) = k/Q$$

where,

$$k=(2*p*(2m* D)1/2/h \quad Q=(2*p*(2m*V)1/2 /h$$

and: V—well depth; L—well width; m*—electron effective mass; h—Planck constant.

Unfortunately, in the real-life systems we cannot obtain perfect separation of adjacent wells and this results in coupling of their wave functions. Thus, instead of discrete levels, this yields us subbands. An equal width, a, of well and barrier layer is assumed. Two main features common for other superlattice structures are easily seen when a is decreased. First, an energy distance between ground state and subsequent bound states increases, which happens because the well width is getting smaller. Second, the discrete levels split into the subbands as a result of thinner barrier layers and poorer separation of neighbouring wells. At this point, it is also important to notice, that the energy discontinuity, i.e. well depth, cannot be simply derived from such material constants as an energy gap, the conduction and valence band structure etc. in each semiconductor because of the lack of knowledge of the common energy reference. This parameter must be either calculated or measured.

Semiconductor Wafer: A tiny complex of electronic components and their connections, produced in or on a small slice of material (like silicon). For more information see Semiconductor manufacturing: How a chip is made (Texas Instruments, USA).

Semiconductor: (1) Nonmetallic material, such as silicon, whose electrical conductivity is moderate and alterable by doping. (2) A pure semiconductor is a poor electrical conductor but when certain impurities are added, the conductivity can increase by orders of magnitude. Common semiconductors are silicon, germanium, and gallium-arsenide. Computer chips are usually made from thin wafers cut from large single crystals of silicon. The silicon is a badly conducting matrix in which better conducting regions are defined by adding impurity atoms (often called dopants). The conductivity of a semiconductor can be described by the formula ó = *ne*ì, where ó is the conductivity measured in units of Ohm^{-1} m^{-1}, *n* is the density of free charge carriers measured in m^{-3}, *e* is the electron charge -1.6022 × 10^{-19} C, and ì is the mobility of the semiconductor measured in m^2 V^{-1} s^{-1}. In a doped semiconductor, the density of free charge carriers is typically equal to the density of dopant atoms that have been added. Thus, the conductivity depends on the number of dopant atoms. Each dopant atom contributes one charge carrier. Silicon is the second most common element in the earth's crust (the first most common is oxygen).

Silicon is commonly found in sand and stone in the form of silicon-dioxide. A substance or object with conductive properties between those of a conductor and a insulator. Material whose conductivity, due to charges of both signs, is normally in the range between that of metals and insulators and in which the electric charge carrier density can be changed by external means.

Material whose conductivity, due to charges of both signs, is normally in the range between that of metals and insulators and in which the electric charge carrier density can be changed by external means.

Sensitivity: (1) 100% sensitivity = 100% true positives, 0% false positives. (2) The amount of change in a sensor's output in response to a change at a sensor's input over the sensor's entire range. Provides an indication of a sensor's ability to detect changes. For some sensors, the sensitivity is defined as the input parameter change required to produce a standardized output change. (3) The minimum input signal required to produce a measurable output signal having a specified signal-to-noise ratio. The amount of change in a sensor's output in response to a change at a sensor's input over the sensor's entire range.

Sensor Model Language SensorML: *Sensor Model Language (SensorML) for In-situ and Remote Sensors*. Edited by Mike Botts (University of Alabama in Huntsville). Open GIS Discussion Paper [not an adopted standard]. Open GIS Consortium Inc. Issued by the OGC Natural Resources and Environment (NRE) Working Group.

Sensor Web: An independent network of wireless, intra-communicating sensor pods, deployed to monitor and explore a limitless range of environments. This adaptable instrument can be tailored to whatever conditions it is sent to observe.

Sensor: A device that responds to the absolute value of, or change in, a physical stimulus (heat, light, sound, magnetism, pressure, or particular motion) and produces a corresponding signal. It may react also to other stimuli, e.g. temperature. An equipment which detects, and may indicate, and/or record objects and activities by means of energy or particles emitted, reflected, or modified by objects.

Sentience Quotient: In the article "Xenopsychology" by Robert Freitas in Analog of April 1984 there is an interesting index called "Sentience quotient". It is based on: The sentience of an intelligence is roughly directly related to the amount of data it can process per unit time and inversely to the overall mass needed to do that processing. This would be something like baud/kilograms. And since that would rapidly turn into a real big number, base 10 logs are used. The "least sentient" would be one bit over the lifetime of the universe massing the entire known universe, or about -70. The "most sentient" is claimed to be +50. Homo sapiens are around +13, a Cray I is +9, a venus flytrap is a peak of +1 with plants generally -2.

Sequencing: (proteins, nucleic acids) Analytical procedures for the determination of the order of amino acids in a polypeptide chain or of nucleotides in a DNA or RNA molecule. [IUPAC Compendium] Sequencing glossary

Serine/threonine Kinase: Protein kinase that phosphorylates serines or threonines on its target protein.

Sertoli Cell: Supporting cell of the mammalian testis that surrounds and nourishes developing sperm cells.

SEU (Single-Event Upset): An IC failure mechanism that occurs when ICs are exposed to cosmic rays A single cosmic ray can bring about latch-up in a CMOS structure by activating a parasitic transistor.

Sex Chromosome: Chromosome that may be present or absent, or present in a variable number of copies, according to the sex of the individual; in mammals, the X and Y chromosomes.

Sexy Technologies: What makes technologies sexy? It seems to be a combination of being new, innovative, challenging, affording clever people a chance to learn new skills (and demonstrate how competitive and bright they are) and expensive (or otherwise unavailable to everyone). A quick search of the web identifies high- speed computers, robotics, nanotechnology, HDTV, Java, wireless communications and biomaterials as "sexy" by some criteria. I'd be interested to hear other interpretations and nuances of this class of technologies. Are there significant differences in what are sexy technologies to biologists, businesspeople, chemists, computer scientists and others? Business of biopharmaceuticals glossary

Shape Memory Alloys: (SMA's) are a unique class of alloys which are able to "remember" their shape and are able to return to that shape even after being bent. The ability is known as the shape memory effect. ... This property has lead to many uses of SMA from orthodontics and coffee makers to methods of controlling aircraft and protecting buildings from earthquake damage. ... The first SMA to be discovered and the most commonly used is called Nitinol. Alloys that can be deformed by up to 8% strain. They will stay in the deformed shape until heated to above the transformation temperature of the specific alloy and then will return to their predeformed shapes. Increases the degrees of freedom of movement in endoscopes used for minimally invasive surgery.

Shape of Crystal: The atoms in a crystal occupy positions with definite geometrical relationships to each other. This structural arrangement of its atoms is uniquely defined by the chemistry of the substance and determines the shape of the crystal.

Shear Modulus: Shear stress divided by shear strain; has the units of force per unit area.

Shear: A shear deformation is one that displaces successive layers of a material transversely with respect to one another, like a crooked stack of cards. Shear is a dimensionless quantity measured by the ratio of the transverse displacement to the thickness over which it occurs.

Sheet Resistance: A characteristic of material that is formed in such a way so that its resistivity varies as a function of its cross-sectional structure. Sheet resistance is expressed as RS in terms of ohms per square, where the square refers to surface area but has no absolute dimensions.

Shim: A thin tapered piece of material used to fill fissures or make a structure flat.

SI (International System of Units): This system is

Canada's official system of measurement. The metre and the Hertz are among the units used in this system.

SI Units: International System of units based on the metric system and units derived from the metric system.

Si/Pyrex Sandwich: An anodic bonding technique used to create novel MEMS devices and also as a batch-compatible packaging process.

Side-Brazed Ceramic Package: A multilayer ceramic dual in-line package (DIP) with external leads brazed on the sides of the package.

Sigma Bond: A *covalent bond* in which *overlap* between two atomic *orbitals* (e.g., of *sp, sp2, or sp3* hybridization) produces a single bonding orbital in which the distribution of shared electrons has a roughly cylindrical symmetry about the axis linking the two atoms; see *pi bond*, *single bond*, *double bond*, *triple bond*. By themselves, sigma bonds present little barrier to rotation of one substructure with respect to another, although *steric* effects and *cyclic* structures may hinder or block rotation.

Signal to Noise: Ratio which can interfere with detection. Can also refer to data analysis. Biological data is often very "noisy". This is particularly seen when trying to look at low abundance *biomolecules.*

Signal-Conditioning Electronics: Electronics that interface a wide range of sensors and transmitters that carry information regarding temperature, humidity, or pressure, to data storage devices.

Signals: Produced by dyes, fluorescence or radioactivity. (Non- radioactive materials, because of disposal and other problems, and improvements in other technologies) are increasingly in demand.

Signal-to-noise-ratio: The ratio of the output signal with an input signal to the output signal with no input signal. A measure of signal strength relative to background noise. The ratio is usually measured in decibels (dBs). Communications engineers always strive to maximize the S/N ratio.

Silane (SiH_4): A toxic gas used to deposit silicon films in CVD processes.

Silica-On-Silicon Waveguide Technology: A technology used to fabricate advanced photonic devices on a small chip using a technique similar to fabricating IC chips.

Silicide: A compound of metal and silicon. Silicides may be easily formed by thermally reacting a variety of metals with silicon and are widely used as contacts to silicon and conductors.

Silicon Fusion Bonding: High-temperature hermetic bonding. Its atomic nature means high-quality bond interfaces.

Silicon Nitride (Si_3N_4): An insulator with a dielectric constant of 7.5 and is used as a barrier material on the sides of the gate as protection against contaminants.

Silicon Sculpting: The act of "etching" bulk silicon. It is one of the basic technologies utilized in traditional bulk micromachining processes.

Silicon: A pure crystalline element (Si) that is the most commonly used semiconductor material for the fabrication of ICs. Since silicon is the only semiconductor that allows growth of its oxide easily, silicon is used in the IC industry and not germanium, which has better semiconductor properties than silicon.

Silicon-Dioxide (SiO_2): An insulator with a dielectric constant of 3.9.

Silicon-Gate MOS: MOS devices with a polycrystalline silicon layer deposited to form the gate elements.

SIMOX (Separation by Implantation of Oxygen): A popular technique used to fabricate silicon-on-insulator (SOI) substrates by implanting oxygen ions at extremely high temperatures in silicon wafers to form oxide layers. Developed by IBM, the relatively perfect silicon oxide allows direct bonding to the pure crystalline silicon substrate.

Simulation: The use of a CAD tool that mimics the behavior of the actual circuit through a collection of sub-circuit models and so enables designers to test and debug circuits before layout and fabrication.

Single Bond: A sigma bond having no associated pi bonds.

Single Cell Repair Unit: A cell repair unit using cilia for propulsion and equipped with a nanocomputer having 10 megabytes of fast RAM and 1 gigabyte of slower-access memory. The unit is extending 1000 individually-controlled molecular manipulators.

Single Crystal: A crystalline solid for which the periodic and repeated atomic pattern extends throughout its entirety without interruption.

Single Electron Devices (SED): Nanoscale devices that control the movement of individual electrons, may one day make it possible for integrated circuits to have as many as 10 billion electronic devices in a square centimeter, a density 1000 times greater than that believed feasible for conventional integrated circuits. In development since the mid-1980s, these devices consist of two electrodes (typically 30 nm wide) separated by a 1 nm- deep insulating layer through which single electrons can tunnel. These devices have many potential applications, from building more sensitive measurement devices to understanding fundamental problems in physics. In the last several years, researchers have built two- junction devices that share a middle electrode. These devices are called "*single- electron transistors*," because, like conventional transistors, their current can be controlled by modifying the surface charge on the middle electrode, making it an ideal element for an integrated circuit. A circuit made of single- electron devices, however, would have to be operated at a temperature of 4 K or below to reduce thermal effects which disturb the movements of single electrons in the solid.

Single Electron Transfer : Exchange of one electron at a time between two electrodes.

Single Foglet: "Imagine a microscopic robot. It has a body about the size of a human cell and 12 arms sticking out in all directions. A bucketfull of such robots might form a 'robot crystal' by linking their arms up into a lattice structure. Now take a room, with people, furniture, and other objects in it — it's still mostly empty air. Fill the air completely full of robots. The robots are called Foglets and the substance they form is Utility Fog, which may have many useful medical applications. And when a number of utility foglets hold hands with their neighbors, they form a reconfigurable array of 'smart matter.'" Copyright Dr. J. Storrs Hall Research Fellow of the Institute for Molecular Manufacturing. "Here's a short list of the powers you'd have or appear to have if embedded in fog: Creation—causing objects to appear and disappear on command. Levitation—causing objects to hover and fly around. Manipulation—causing forces (squeezing, hitting, pulling) on objects (real ones) at a distance. Teleportation—nearly any combination of telepresence and virtual reality between fog-filled locations."

Single Molecule Detection: Recent advances in optical imaging and biomechanical techniques have demonstrated that it is possible to make observations on the dynamic behavior of single molecules, to determine mechanisms of action at the level of an individual molecule, and to explore heterogeneity among different molecules within a population. These studies have the potential to provide fundamentally new information about biological processes and are critical for a better understanding of cellular function. ... Single molecule methods are likely to lead to significant advances in understanding molecular movement, dynamics, and function.

Single-Charge Electronics: Even though the study of single electron charging effects with granular metallic systems dates back to the 1950s, it was the research of Likharev and coworkers almost 10 years ago that laid much of the groundwork for understanding single-charge transport in nanoscale tunnel junctions. The concept was developed of a Coulomb gap that can be exploited to control the transfer of single charges. Since then, many research groups have made use of the Coulomb blockade effect to develop SED technology. through 50 show some of the myriad approaches to developing SEDs and representative laboratories pursuing the various SED concepts. Some of the more recent results. The group at Hitachi Europe uses a side-gated constriction in a delta doped GaAs structure to fabricate a magnetic tunnel junction device in which a series of small islands separated by tunnel barriers are formed. At ~ 2 K, the Coulomb gap voltage oscillates as a function of the side-gate voltage. Using the MTJ device as a building block, both memory and logic (inverter, NOR) functions have been demonstrated. The fabrication procedure makes use of standard semiconductor processing techniques and does not rely on lithography to define the nanoscale islands, since these are created by disorder in the delta-doped layer.

Single-Crystal Silicon: Silicon having a continuous, regular crystalline lattice structure and having no internal grain boundaries.

Single-Electron Device (SED): A nanoscale device that controls the movement of individual electrons. One day, it may be possible for ICs to have as many as 10 billion electronic devices in a square centimeter, a density 1000 times greater than that believed feasible for single-electron transistors.

Single-electron Transistor (SET): SET is a three terminal device, where single electron current between a source and a drain through a nanocrystal is controlled by a gate. The nanocrystals are the tiny light specs.

Single-Mode Fiber: An optical fiber in which the signal travels in one mode. It typically has an 8-10-mm core within a 125-mm cladding.

Single-Mode Fiber-Optical Switch: An optical switch in which only the lowest-order bound mode can propagate at the wavelength of interest. The bound mode refers to a mode that has a field intensity that decays monotonically in the transverse direction, everywhere external to the core. A mode is one of the various possible patterns of propagating or standing electromagnetic fields, which is characterized by wavelength, frequency, and electric and magnetic field strengths.

Single-pair FRET spFRET: Designed to overcome the averaging effects of ensemble studies because measurements are made on single molecules freely diffusing in solution. This method limits the observation period to the diffusion time of each molecule through the focal spot of a laser on the order of a few hundred milliseconds, but it permits the rapid gathering of data at single- molecule resolution on a large number of molecules in a short time period. SpFRET can be used to study intramolecular conformational changes by placing the donor and acceptor fluorescent tags on two different sites of the same macromolecule, or alternatively, intermolecular interactions can be studied by attaching the donor and acceptor tags to two different macromolecules.

Single-Point Prediction: A forecast in the form of a single value for a given market and for a given period of time. It is usually obtained by taking the average of a series of expert opinions.

Singlet: An electronic state of a molecule in which all spins are paired; see doublet, triplet.

Singularity: Defined by Vernor Vinge as the "postulated point or short period in our future when our self-guided evolutionary development accelerates enormously (powered by nanotechnology, neuroscience, AI, and perhaps uploading) so that nothing beyond that time can reliably be conceived. ...a future time when societal, scientific and economic change is so fast we cannot even imagine what will happen from our present perspective, and when humanity will become posthumanity." Another definition is the singular time when technological development will be at its fastest. A grand evolutionary leap.

Sinter: The process of heating a material to just below its melting point. An extended period of sintering is the method by which the constituent components of a ceramic superconductor are combined in a solid-state reaction. Since ceramic superconductors are inherently brittle, sintering helps promote intergranular bonding and hardness.

Sintering: Particle coalescence of a powdered aggregate by diffusion that is accomplished by firing at an elevated temperature.

SIP (Single In-Line Packaging): Packaging with a single row of pins (in picket-fence style) and is usually mounted in the vertical plane.

Sky Hook: A long, very strong, cable in orbit around a planet which rotates around its center of mass in such a way that when one end is closest to the ground, its relative velocity is almost zero. It would function as a kind of space elevator; shuttle craft would anchor to the end and then be lifted into orbit where they would be released. It is closely related to the idea of a beanstalk.

Slingshot: The process by which many space exploration

satellites and probes get into deep space. A space exploration satellite can enter the gravitational field of a large planetary body like the Earth or Mars, and use the force of that planet's gravity to give the satellite momentum. Basically, the satellite swings around a planet and uses the energy from the planet's gravitational pull to send it into space. This uses the planet's gravitational pull like a slingshot.

Slip Casting: A forming technique used to shape ceramic materials. A slip or suspension of solid particles in water is poured into a porous mold. A solid layer forms on the inside wall as water is absorbed by the mold, leaving a shell (or a solid piece) in the shape of the mold.

Slip: Plastic deformation resulting from dislocation motion; also, the shear displacement of 2 adjacent planes of atoms.

Small Asteroid Retrieval: In situ resource utilization is undoubtedly necessary for large scale colonization of the solar system. Asteroids are particularly promising for orbital use since many are in near Earth orbits. Moving asteroids into low Earth orbit for utilization poses a safety problem should the asteroid get out of control and enter the atmosphere. Very small asteroids can cause significant destruction. The 1908 Tunguska explosion, which was calculated to be a 60 meter diameter stony asteroid, leveled 2,200 km^2 of forest. Hills calculated that 4 meter diameter iron asteroids are near the threshold for ground damage. Both these calculations assumed high collision speeds. At a density of 7.7 g/cm^3 Babadzhanov, a 3 meter diameter asteroid should have a mass of about 110 tons. Rabinowitz estimates that there are about one billion ten meter diameter near Earth asteroids and there should be far more smaller objects. For colonization applications one would ideally provide the same radiation protection available on Earth. Each square meter on Earth is protected by about 10 tons of atmosphere. Therefore, structures orbiting below the van Allen belts would like 10 tons/meter^2 surface area shielding mass. This would dominate the mass requirements of any system and require one small asteroid for each 11 meter^2 of colony exterior surface area. A 10,000 person cylindrical space colony such as Lewis One with a diameter of almost 500 meters and a length of nearly 2000 meters would require a minimum of about 90,000 retrieval missions to provide the shielding mass. The large number of missions required suggests that a fully automated, replicating nanotechnology may be essential to build large low Earth orbit colonies from small asteroids. A nanotechnology swarm along with an atomically precise lightsail is a promising small asteroid retrieval system. Lightsail propulsion insures that no mass will be lost as reaction mass. The swarm can control the lightsail by shifting mass. When a target asteroid is found, the swarm spreads out over the surface to form a bag. The interface to the sail must be active to account for the rotation of the asteroid—which is unlikely to have an axis-of-rotation in the proper direction to apply thrust for the return to Earth orbit. The active interface is simply swarm elements that transfer between each other to allow the sail to stay in the proper orientation. Of course, there are many other possibilities for nanotechnology based retrieval vehicles

Small Molecule Therapeutics: Low molecular-weight drug. Compared to larger molecular weight pharmaceuticals such as proteins, peptides, and carbohydrates, small molecules can more easily penetrate cell membranes and the blood brain barrier. Can be delivered orally or intravenously. These molecules tend to incur lower process development and manufacturing costs.

Small Molecules: Preferred for drugs as they are orally available (unlike proteins which must be administered by injection or topically). Size of small molecules is generally under 1000 Daltons, but many estimates seem to range between 300 to 700 Daltons.

Smart Dust: Given a mature nanotechnology, it should be possible to build sensors in balloon-borne systems approximately the size of bacteria. With replication based manufacturing, these should be quite inexpensive. If the serious communication and control problems can be solved, one can imagine spreading billions of tiny lighter-than-air vehicles into the atmosphere to measure wind currents and atmospheric composition. A similar approach might be taken in the oceans—note that the oceans are full of floating microscopic living organisms that can sense and react to their environment. Smart dust might sense the environment, note the location via a GPS-like system, and store that information until close enough to a data-collection point to transfer the data to the outside world.

Originally, the name given to a research project at Berkeley University. The goal of the project was to create cubic millimeter devices integrating a complete sensor/communication system (sensor, power supply, analog circuitry, bidirectional optical communication, and a programmable microprocessor). Based on MEMS technology, these "macro motes" were designed to fit multiple applications, including Defense-related sensor networks or temperature and humidity monitoring... Today, the expression is often used with a wider meaning, and refers to micro or nano-sized components that could be used in numbers and perform all kinds of monitoring.

Smart Materials: Here, materials and products capable of relatively complex behavior due to the incorporation of nanocomputers and nanomachines. Also used for products having some ability to respond to the environment. If you combined microscopic motors, gears, levers, bearing, plates, sensors, power and communication cables, etc., with powerful microscopic computers, you have the makings of a new class of materials: "smart materials." Programmable smart materials could shape-shift into just about any desired object. A house made of smart materials would be quite useful and interesting. Imagine a wall changing color at your command, or making a window where their was none before.

Several different types of materials exhibit sensing and actuation capabilities, including ferroelectrics (exhibiting strain in response to a electric field), shape- memory alloys (exhibiting phase transition- driven shape change in response to temperature change), and magnetostrictive materials (exhibiting strain in response to a magnetic field). These effects also work in reverse, so that these materials, separately or together, can be used to combine sensing and actuation in response to environmental conditions. They are currently in widespread use in applications from ink-jet printers to magnetic disk drives to anti- coagulant devices. An important class of smart materials is composites based upon lead zirconate titanate (PZT) and related ferroelectric materials that allow increased sensitivity, multiple frequency response, and variable frequency.

Smart Polymers: Ionic gels that deform in response to electric fields. Such electro- active polymers have already been used to make "artificial muscles" (Shahinpoor et al., 1998). Currently available materials have limited mechanical power, but this is an active research area with potential applications to robots for space exploration, hazardous duty of various types, and surveillance. Hydrogels that swell and shrink in response to changes in pH or temperature are another possibility; these hydrogels could be used to deliver encapsulated drugs in response to changes in body chemistry.

Smart Sensor: A sensor in which the electronics that process the output from the sensor, and forms the modifier, are partially or fully integrated on a single chip.

Smart Switches: Optical switches that can provide multiple functions such as switching, power regulation and dynamic broadcast. The integration of power regulation and switching reduces complexity of optical-network control and management.

Smartdust: Also "Smartdust Motes" "...tiny, bottle-cap-shaped micro-machines fitted with wireless communication devices—that measure light and temperature [among other things, such as environmental monitoring, health, security, distributed processing and tracking—ed]. When clustered together, they automatically create highly flexible, low-power networks with applications ranging from climate-control systems to entertainment devices that interact with handheld computers."

SMD (Surface-Mount Device): A sophisticated microelectronic packaging technology in which packages are small enough to be soldered using eutectic alloys on the surface of a printed-circuit board.

SNP (Single Nucleotide Polymorphism): Polymorphism differing in a single base pair (AT or GC).

SNP Genotyping: A method of determining variations in genetic sequences that cause diseases. SNP is pronounced as "snip,"

SNT: An abbreviation for structural nanotechnology; refers to integration of nanotech features into non-MNT products, also called nanomaterials.

SO/SOIC (Small-Outline/Small-Outline IC): A plastic dual in-line package with "gull wing" feet designed for surface mounting. Most versions have lead spacing compatible with that of PLCC packages. Another version, SOT (Small Outline Transistor) has been popular in hybrid use for several years.

SOA (Semiconductor Optical Amplifier): It is similar to regular lasers, but with an optical signal being amplified by stimulating light emission. It is manufactured by coating an optical cavity that stores light, by coating it with anti-reflection coatings and angle cleaving its facets. An SOA provides high-speed switching capability as well as gain.

Soft Bake: A bake performed after photoresist coating and before exposure. This step dries the photoresist film and improves photoresist adhesion to the wafer. It is performed at a temperature lower than a hard bake.

Soft Lithography: A method for transferring a surface pattern that makes use of an elastomeric stamp.

SOI: (Silicon On Insulator): An oxide layer sandwiched between two silicon layers, used in etching, as an etch stop barrier and sacrificial layer.

Solar Max: A NASA space exploration satellite that studied the sun. In 1984, when the satellite's instruments failed, astronauts on board the shuttle repaired it after first grabbing it with the Canadarm.

Solar Panel: A source of power for satellites, made up of many individual solar cells, which are devices for transforming the energy of sunlight into electricity.

Solid State: Refers to electronic devices, like transistors, that conduct and control the movement of electrons in solid materials (instead of in a vacuum).

Solid-State Imaging Instrument (SSI): A camera currently in use on the space exploration satellite Galileo.

Solidus Temperature: In a mixture of materials, it is the temperature below where the mixture starts to solidify completely and below where it cannot liquefy. In a phase diagram it is the line between the solid state and the intermediary state between solid and liquid.

Solution Arrays: Functionalized particle sets for multiplexed bioanalysis in solution.

Solvent: The component of a solution that dissolves a solute.

SONET (Synchronous Optical Network): Communication between different networks requires complicated multiplexing/demultiplexing and coding/decoding processes to convert a signal from one format to another. The SONET communication protocol allows these different types of formats to be transmitted on one line. SONET allows the addition and elimination of signals with a single multiplexer.

SOS (Silicon-On-Sapphire): A fast MOS technology in which silicon is epitaxially grown on a sapphire wafer, and etched away between transistors. Each device is thus isolated by air or oxide from other devices.

Sound Power Level (dB): Equal to 10 x $\log_{10}$ (sound power of the source /a reference sound power). Sound power values are used to find out whether sound levels are not hazardously high.

Sounder: A special kind of radiometer which measures changes of atmospheric temperature and changes in water vapor content of the air at various heights in the atmosphere.

Source: The working-current terminal of an FET that is a source for holes (P-channel) or free electrons (N-channel) flowing in the channel. It corresponds to the emitter of bipolar transistors.

***sp, sp2, sp3*:** An isolated carbon atom has four *valence orbitals*: three mutually perpendicular *p* orbitals, each with a single nodal plane, and one spherically symmetric *s* orbital. A carbon atom in a typical molecule can be regarded as bonding with four orbitals consisting of weighted sums (termed hybrids) of these *s* and *p* orbitals. One common pattern has four equivalent orbitals, each formed by combining the three *p* orbitals with the *s* orbital; this is *sp3* hybridization. An *sp3* carbon atom forms four *sigma bonds*, usually in a roughly tetrahedral arrangement. Another common pattern has three equivalent orbitals formed by combining two *p* orbitals with the *s* orbital; this is termed *sp2* hybridization. An *sp2* carbon atom forms three roughly coplanar sigma bonds, usually separated by ~120 , and one *pi bond* (or several fractional pi bonds). If a single *p* orbital is combined with the *s* orbital, the result is *sp* hybridization, forming two sigma bonds and two pi bonds (usually in a straight line). Atoms of other kinds (e.g., N and O) can hybridize in an analogous manner.

Space Elevator: Issacs and Pearson proposed a space elevator—a cable extending from the Earth's surface into space with a center of mass at geosynchronous altitude. If such a system could be built, it should be mechanically stable and vehicles could ascend and descend along the cable at almost any reasonable speed using electric power (actually generating power on the way down). The first incredibly difficult problem with building a space elevator is strength of materials. Maximum stress is at geosynchronous altitude so the cable must be thickest there and taper exponentially as it approaches Earth. Any potential material may be characterized by the taper factor—the ratio between the cable's radius at geosynchronous altitude and at the Earth's surface. For steel the taper factor is tens of thousands—clearly impossible. For diamond, the taper factor is 21.9 including a safety factor. Diamond is, however, brittle. Carbon nanotubes have a strength in tension similar to diamond, but bundles of these nanometer-scale radius tubes shouldn't propagate cracks nearly as well as the diamond tetrahedral lattice. Thus, if the considerable problems of developing a molecular nanotechnology capable of making nearly perfect carbon nanotube systems approximately 70,000 kilometers long can be overcome, the first serious problem of a transportation system capable of truly large scale transfers of mass to orbit can be solved. The next immense problem with space elevators is safety—how to avoid dropping thousands of kilometers of cable on Earth if the cable breaks. Active materials may help by monitoring and repairing small flaws in the cable and/or detecting a major failure and disassembling the cable into small elements

Space Exploration Satellite: Technically a space probe because it is sent deep into space and does not necessarily orbit anything. These are not satellites because the definition of a satellite is something that is in orbit around something else. However, space exploration satellites are similar to orbiting satellites in design and function. Like astronomy satellites, they study faraway planets and other stellar phenomena.

Space Fountain: A vertical stream of magnetically accelerated pellets reaching out into space, where a station held aloft by its momentum reverses the direction and directs it towards a receiver on the ground. Essentially a simpler version of a Lofstrom loop.

Space Telescope Imaging Spectrograph (STIS): A special spectrograph found on the Hubble Space Telescope.

Span: The modulus of the difference between the two limits of a nominal range of a measuring instrument. E.g., a nominal range –10 volts to +10 volts spans 20 volts. The difference between the highest and lowest scale values of an instrument.

Spark Erosion: A method for producing fine powders of metals, alloys, and semiconductors by maintaining cyclic spark discharges near material immersed in a dielectric liquid. The spark discharge creates localized vaporization of the material in the dielectric, which freezes into powder. Spark erosion is an electro-discharge machining technique used to remove material during the manufacture of tools and dies.

Special Cause Variation: A variation in output for a process because a component of the process changed.

Species: In chemistry, a distinct kind of molecule, ion, or other structure.

Specific Heat: The heat capacity per unit mass of a given material. It is the heat required to raise the temperature of one gram of a substance one degree Celsius.

Specific Modulus (specific stiffness): The ratio of elastic modulus to specific gravity for a material.

Specific Strength: The ration of tensile strength to specific gravity for a material.

Specificity: In the five disciplines that eventually contributed to the formation of molecular biology, ideas of specificity had widely different standing and character. In microbiology, the relevance of specificity in the present day sense awaited resolution of the long, piecemeal shift of opinions about what sort of creatures micro- organisms were....[Oswald Avery, Chargaff, Lwoff, Luria Delbruck, Jacques Monod, Alice Audureau] Mendelian genetics itself had always been highly specific, of course, and from very early the genetic specificity—the map—was understood to form a strictly linear array [Thomas Hunt Morgan, Boris Ephrussi, George Beadle, Edward Tatum]... In physical chemistry, specificity was not abstract and not linear, but concretely physical and three-dimensional [Pauling] ... Crystallography, in its way, was also permeated with specificity [Edward Tyson Reichert, Amos Peaslee Brown, Felix Haurowitz] ... Specificity in the present- day sense of unique molecular structures was never a surprise to the crystallographers ... The fifth discipline in the synthesis that formed molecular biology was biochemistry. Yet the standard view of the rise of molecular biology has somewhat taken for granted, for example, the radical sharpening of ideas of specificity represented by Fritz Lipmann's elucidation of the way energy is supplied to the steps of cellular reactions. And it grievously undervalues the work of the two biochemists who proved decisive in changing the way people thought about specificity. The man who released the present day understanding of molecular specificity in living processes was Frederick Sanger ... the most general and profound [result of his methods] was that proteins are entirely and uniquely specified. 100% specificity = 100% true negatives, 0% false negatives.

Spectrograph: An instrument that spreads out the light gathered by a telescope so that it can be analyzed to determine many different properties of celestial objects. These properties include the chemical composition and abundance of different elements, temperature, radial velocity, rotational velocity, and magnetic fields.

Spectroscopic Dyes: These dyes—in particular, nanoparticles — are emerging as alternatives to fluorescent dyes. Because the emission spectra of nanoparticles vary according to these particles' specific size and shape, these nanostructures can be used in multicolor detection formats, potentially offering much greater multiplexing than is currently achievable. The fact that chemists have been able to create a great variety of structures (and properties) in nanoparticles suggests that these particles might be more "finely tunable" than organic dyes, allowing better results from biological assays CHI High-Content Analysis Market Outlook report, 2004

SPICE (Simulation Program for Integrated Circuits Emphasis): An Electronic Circuit simulator used to verify circuit designs and predict the circuit behavior.

Spider Bonding: A method of connecting an IC die to its package leads. A preformed leadframe is placed over the chip and all contacts are made by just one operation of a bonding surface. Tape automated bonding (TAB) uses this approach for interconnection.

Spike, The: Another term for the singularity, suggested by Damien Broderick since the growth curves look almost like a spike as it is approached. [Damien Broderick, The Spike 1997] [AS]

Spin: A quantum number describing the angular momentum and magnetic moment of a subatomic particle.

Spinning: Fiber forming process: a multitude of fibers are spun as molten material is forced through many small orifices.

Spin-On-Glass (SOG): A thin layer of liquid glass spin-coated on a surface that solidifies to form a dielectric.

Spin-Spin Coupling: Interaction between the spin states of two subatomic particles.

Spin-stabilized: Attitude control is done by having the entire satellite spin. It becomes very stable because the satellite as a whole is acting as a gyroscope.

Spintronics: The technology utilizing the electron spin to manipulate (calculate) or store information. The 'classical' electronics, with respect to spintronics should possibly be called 'chargetronics' since it uses electron charge, i.e. charge currents to manipulate the information (fire the transistors and similar). The spin is the quantum number of an electron which can take two values (states)—'up' and 'down'. In spintronics, the *spin* currents are of relevance to the function of electronic devices. Note that the spin currents are also the charge currents, since the carrier of spin (electron) is also a carrier of charge. The materials that would support the spintronical applications would have to be sensitive to the sign (up or down) of the spin currents, i.e. their response should be dependent on the sign of the spin current. Materials containing several layers of different magnetic and nonmagnetic metals that have thicknesses in the range of several nanometers are the candidates for such applications. This, of course, requires material nanostructuring.

Spintronics: Devices that use the electron spin instead of the electron charge for electronics.

SPR Surface Plasmon Resonance: A biosensing technique in which biomolecules capable of binding to specific analytes or ligands are first immobilized on one side of a metallic film. Light is then focused on the opposite side of the film to excite the surface plasmons, that is, the oscillations of free electrons propagating along the film's surface. The refractive index of light reflecting off this surface is measured. When the immobilized biomolecules are bound by their ligands, an alteration in surface plasmons on the opposite side of the film is created which is directly proportional to the change in bound, or adsorbed, mass. Binding is measured by changes in the refractive index. The technique is used to study biomolecular interactions, such as antigen—antibody binding.

Spreading-Resistance Probe (SRP): A metrology technique to measure electrically active dopant levels. A semiconductor is ground down at an angle through the area to be measured and four tiny needles are pushed down measuring resistance as they go.

Sprue And Runner System: A type of mold system used in metallurgy. A sprue is the vertical channel from the top of the mold to the mold cavity where the metal has to reach. The runner is a cylindrical tube that connects the sprue with the riser.

Sputtering: (1) The ejection of atoms or groups of atoms from the surface of the cathode (negative electrode) of a vacuum tube as the result of heavy-ion impact. (2) The use of this process to deposit a thin layer of metal on a glass, plastic, metal, or other surface in a vacuum.

Squeeze-film Damping: Effect of ambient fluid and spacing on the vertical movement of a structural member with respect to a substrate.

SQUID: A superconducting loop interrupted in 2 places by Josephson junctions. When sufficient electrical current is conducted across the squid body, a voltage is generated proportional to the strength of any nearby magnetic field. The SQUID, an acronym for Superconducting QUantum Interference Device, is the most sensitive detector known to science.

SRAM (Static Random-Access Memory): Memory circuits that permit the storage and retrieval of data from any desired memory cell. The data remain stored as long as power is applied. It does not require periodic refresh cycles, as does dynamic RAM.

Stability: The ability of a sensor to retain specified characteristics after being subjected to designated environmental or electrical test conditions.

Stabilizer: A polymer additive that counteracts deteriorative processes.

Stable: Strictly speaking, a system is termed stable if no rearrangement of its parts can form a system of lower *free energy*. In practice, the term is used with an implicit proviso regarding the transformations to be considered. Hydrogen is not considered unstable merely because it is subject to nuclear fusion at extreme temperatures. A system is usually regarded as stable (more precisely, as kinetically stable) if its rate of transformation to a state of lower free energy is negligible (by some standard) under the ambient conditions. In *nanomechanical* systems, a structure can commonly be regarded as stable if it has an extremely low rate of transformations when subjected to its intended operating conditions.

Staining and Labeling: The marking of biological material with a dye or other reagent for the purpose of identifying and quantitating components of tissues, cells or their extracts. [MeSH, 2001]

Standard Atmospheric Pressure: A unit of pressure equal to the earth's normal air pressure at an altitude of sea level, i.e. 1.01325×10^5 Newton per square meter.

Standard Cell: A pre-designed IC module that implements a specific function. The data required to use the cell as part of a more complex circuit are contained in a computer file that can be accessed by other CAD tools. Standard cells are designed to be interconnected by simple abutment with other cells from the same cell library.

Standard Deviation: The standard deviation for a series of n measurements characterizes the dispersion among them and is given by the formula

Standard Hydrogen Electrode (SHE): A platinum conductor in contact with a 1 M H^+ ions and bathed by hydrogen gas at one atmosphere.

Standing Waves: A problem associated with optical lithography and created when two sine-wave-shaped waves having the same frequency and the same amplitude are traveling in opposite directions. Standing waves are created when incoming light waves of the optical system can interfere with the light reflected due to the different refractive indices of photoresist and substrate material. Standing waves create line-width variations and degenerate image quality. Antireflective coatings are used to eliminate it.

Star Trek Scenario: Someone builds potentially dangerous self-replicating devices that spread disastrously.

State: A physical system is said to be in a particular state when its physical properties fall within some particular range; the boundaries of the range defining a state depend on the problem under consideration. In a *classical* world, each point in *phase space* could be said to correspond to a distinct state. In the real world, time-invariant systems in *quantum mechanics* have a set of discrete states, particular superpositions of which constitute complete descriptions of the system. In practice, broader boundaries are usually drawn. A molecule is often said to be in a particular excited *electronic* state, regardless of its state of *mechanical* vibration. In *nanomechanical* systems, the *PES* often corresponds to a set of distinct *potential wells*, and all points in configuration space within a particular well can be regarded as one state. Definitions of state in the thermodynamics of bulk matter are analogous, but extremely coarse by these standards.

Statistical Mechanics: Statistical mechanics treats the detailed *state* of a system (its *quantum* state or, in *classical* models, its position in *phase space*) as unknown and subject to statistical uncertainties; *entropy* is a measure of this uncertainty. Statistical mechanics describes the distribution of states in an *equilibrium* system at a given *temperature* (describing either the distribution of probabilities of quantum states or the *probability density function* in *phase space*), and can be used to derive *thermodynamic* properties from properties at the molecular level. These equilibrium results are useful in *nanomechanical* design.

Statistical Process Control (SPC): A method to control product quality by observing a process during its operation while the products are being produced, rather than depending on inspection. This identifies the causes of variation in product quality.

Stator: Amotionless part in a machine about which a rotor revolves.

Steady-state Diffusion: The diffusion condition for which there is no net accumulation or depletion of diffusing species. The diffusion flux is independent of time.

STEM Antennas: An acronym for Storable Tubular Extendible Member. STEM antennas are stored in the body of a satellite to save space at launch and unrolled when the satellite reaches orbit. Canada's first satellite, Alouette, used four STEM antennas.

Stem Cell Research: Stem cell research is a relatively new technology that uses human embryo cells to treat disease. The embryo cells are called "stem cells" because they can be stimulated to develop into any type of body cell or system. Human embryo cells are especially valuable because they can be used to treat and possibly cure many diseases such as heart disease, diabetes, and Alzheimers. Stem cells have been identified in placental tissue, blood, and fat but these sources are limited in their efficacy. Embryonic stem cells have much more potential and are harvested when the embryos are no more than a few days old. The process destroys the embryos.The biggest source of human embryos is those left over from in-vitro fertilization procedures. Fertility treatments usually result in multiple embryos. Since only the most viable are selected for implantation, the remaining embryos are frozen.

Costs are incurred with the cryogenic (freezing) process and with ongoing maintenance. There is a great deal of controversy over the ethics of stem cell research. Some say that life begins at conception and that the use of humans (even immature ones) for research purposes is immoral. Others claim that the embryos are only tiny amounts of undifferentiated tissue and since they are already scheduled for destruction, they should be used to help others. Adoption programs have resulted in the use of some of these excess embryos and live births have occurred. Since these leftover embryos are typically less viable, they may not result in healthy children. Stem cell research using these leftover embryos or embryos created solely for scientific purposes is currently banned in the United States. As an alternative, the government has set up a Federal Registry of 60 stem cell "lines"—cells whose initial embryos have already been destroyed. Not all of the lines have been approved by the National Institutes of Health (NIH) and scientists have complained that their quality could be inferior. Federal funding is available only for research using these lines. Other countries have embraced the idea and embryonic stem cell research is currently being conducted all over the world. Vast financial rewards await the developers of viable disease treatments. Recent polls have shown increased public support for the idea.

Stem cells and Cloning: There have been some rapid advances in stem cell research and cloning technology during the last few years. The Royal Society has provided advice to policy-makers on some of the scientific, medical, legal and ethical aspects of stem cells and cloning. Background information on stem cells and cloning, and details of the Society's work on the issue, can be found here.

Stent: An expandable, metallic (tubular shaped) device that provides structural support for a vessel.

Step and Repeat: A lithographic system that exposes a wafer by exposing an area of the wafer and then stepping to the next adjacent area of the wafer to perform the next exposure. The system steps and exposes repeatedly until the whole wafer is covered with patterns. This system utilizes reticles to create patterns.

Step Coverage: If a layer of material is to be deposited on top of a thin strip of material, it is the ratio of film thickness along the walls of the step to the film thickness at the bottom of the step. LPCVD and PECVD offer excellent step coverage.

Step Response: The response of a system to an instantaneous jump in the input signal.

Stepper: A resist exposure tool based on projection lithography and step-and-repeat wafer exposure procedures. The same as Stepping Aligner.

Stepping Aligner: An optical system that projects (usually 5X or 10X) reticle images on a portion of the photoresist-coated wafer. After exposure, the wafer is stepped to a new position and the exposure process is repeated.

Steric Hindrance: Slowing of the rate of a chemical *reaction* owing to the presence of structures on the *reagents* that mechanically interfere with the motions associated with the reaction, typically by obstructing the reaction site.

Steric: Pertaining to the spatial relationships of atoms in a molecular structure, and in particular, to the space-filling properties of a molecule. If molecules were rigid and had hard surfaces, steric properties would merely be an opaque way of saying "shape"; a flexible side-chain, however, has definite steric properties but no fixed shape. *Nanomechanical* systems make extensive use of the steric properties of relatively rigid molecules, for which the term "shape" has essentially its conventional meaning so long as one remembers that the surface interactions are soft on small length-scales.

Stewart Platforms: A positional device. John Storrs info and links See also A New Family of Six Degree Of Freedom Positional Devices

Stiction: Undesired adhesion of movable solids in very close proximity, caused by forces such as Van der Waals, capillary, and hydrogen.

Stiffness: The stiffness of a system with respect to a deformation (e.g., the stiffness of a spring with respect to stretching) is the second derivative of the energy with respect to the corresponding displacement; this measures the curvature of the *potential energy surface* along a particular direction. Positive stiffness is associated with stability, and a large stiffness can result in a small positional uncertainty in the presence of thermal excitation. Negative stiffnesses correspond to unstable locations on the potential energy surface. Alternative terms for stiffness include force gradient and rigidity.

STIM (Smart Transducer Interface Module): A module that contains the TEDS logic to implement the transducer interface, the transducers and signal conversion or conditioning units. It contains analog-to-digital or digital-to-analog converters and activating mechanisms.

Stoichiometry: For ionic compounds, the state of having exactly the ratio of cations to anions specified by the chemical formula. Stoichiometric quantities refers to quantities of reactants mixed in exactly the correct amounts so that all are used up at the same time. The quantitative relationship between chemical elements in a chemical substance.

Strain (symbol å): The change in gauge length of a specimen, in the direction of an applied stress, divided by its original gauge length.

Strain Gauge: An element (wire or foil) that measures a strain based on electrical resistance changes of the gauge that result from a change in length or dimension strain of the wire or foil.

Strain: In mechanical engineering, strain is a measure of the deformation resulting from *stress* (that is, force per unit area); the displacement of one point with respect to another, divided by their *equilibrium* separation in the absence of stress. In chemistry, a molecular fragment generally has some equilibrium geometry (bond lengths, interbond angles, etc.) when the rest of the molecular structure does not impose special constraints (e.g., bending bonds to form a small ring). Deviations from this equilibrium geometry are described as strain, and increase the energy of the molecule. Strain in the mechanical engineering sense causes strain in the chemical sense.

Strain: The alteration of relations between the parts of a structure by applying an external force.

Streptavidin: A 60 kD extracellular protein of *Streptomyces avidinii* with four high-affinity biotin binding sites. Unlike AVIDIN, streptavidin has a near neutral isoelectric point and is free of carbohydrate side chains. [MeSH, 1998]

Stress Concentration: High localized stress intensity at a geometric discontinuity of a body, as in the vicinity of a notch or hole or at a change in the diameter of a shaft. The concentration or amplification of an applied stress at the tip of a notch or small crack.

Stress Corrosion (Cracking): A form of failure resulting from the combined action of a tensile stress and a corrosion environment, occurring at lower stress levels than required when the corrosion environment is absent.

Stress: Force per unit area applied by one part of an object to another. Pressure is an isotropic compressive stress. Suspending a mass from a fiber places it in tensile stress. Gluing a layer of rubber between two plates and then sliding one over the other (while holding their separation constant) places the rubber in *shear* stress.

Stripes: Stripes are microscopic rivers of charge that flow across the surface of a Type 2 superconductor. It is theorized that stripes encourage "holes" to pair up and, as such, may play a role in facilitating charge transfer. Recently, at the Stripes 2000 conference in Rome, Italy, it was shown that there exists a critical value of micro-strain that must be exerted upon the CuO_2 planes for stripes to form.

Stripping: The process of completely removing a coating such as photoresist.

Structural Material: Used to make the parts that move in a MEMS device, as opposed to the sacrificial material, which is the material that is etched away to free the structural material.

Structural Volume: The interior of a *diamondoid* structure typically consists of a dense network of *covalent bonds*; a larger *excluded volume*, however, is determined by nonbonded repulsions at the surface. The structural volume corresponds to a region smaller than the excluded volume, chosen to make properties such as the strength and *modulus* nearly size independent by correcting for surface effects.

Structure Activity Relationship (SAR): The relationship between chemical structure and pharmacological activity for a series of compounds. Compounds are often classed together because they have structural characteristics in common including shape, size, stereochemical arrangement, and distribution of functional groups. Other factors contributing to structure- activity relationship include chemical reactivity, electronic effects, resonance, and inductive effects.

Structure Based Design: A design strategy for new chemical entities based on the three- dimensional (3D) structure of the target obtained by X-ray or nuclear magnetic resonance (NMR) studies, or from protein homology models.

Structure Based Drug Design: Structure- Based Drug Design and Virtual Screening bring together high-throughput crystallography and NMR, computational advances in docking algorithms and virtual screening, and traditional techniques of high-throughput screening and combinatorial chemistry to increase the efficiency and speed of the drug discovery and development process. ... the latest developments in de novo, NMR-based, and crystallography-based drug design. For years researchers have sought a more rational approach to designing drugs rather than screening for hits and leads. The rapidly growing body of structural information emerging as a result of genomic- derived targets and industrialization of protein structure determination is dramatically altering the data drug designers have to work with. As a result very encouraging progress is being made in drug design. New approaches, such as co-crystallization of ligands with a given target, allows structural techniques to be used for screening, which then further facilitates the design process. Since the early 1980s, industry has been interested in structural biology as part of the discipline of direct structure- based drug design, which combines structural biology with computational and medicinal chemistry in order to design drugs—rather than merely selecting drugs—that modulate a protein target of interest. CHI Structural proteomics report

SU-8 (Epon SU-8): A photodefinable epoxy used to make high-aspect-ratio features. This material is commonly used in electroplating or polymer casting.

Sublimation: A phase change of a material directly from the solid to the gaseous state without becoming liquid.

Substitutional Atom: An extraneous atom replacing an atom of a crystal and occupying its position in the crystal lattice.

Substrate: The material on which a microelectronic device is built. Such material may be active, like silicon, or passive, like alumina ceramic.

SUMMiT (Sandia Ultra Planar, Multi-level MEMS Technology): A Sandia MEMS fabrication process, consisting of a four-level polycrystalline silicon surface-micromachining process: one ground plane/ electrical interconnect and three mechanical layers. There is also a 5-layer version called SUMMiT V.

Sun-synchronous Orbit: A special case of the polar orbit. In a sun-synchronous orbit, the satellite passes over the same part of the Earth at roughly the same local time each day. Sun-synchronous orbits are usually meduim or low orbits.

Super Critical Drying: A drying technique that avoids the sudden change in densities caused by a fluid's transition from liquid to gas during typical air-drying. Super critical drying is usually performed after release of bulk or surface micromachined devices to achieve stictionless movement by avoiding distortion due to thermal expansion.

Superconductivity: (1) A phenomenon characterized by the disappearance of the electrical resistivity at temperatures approaching 0 K. (2) About one third of all metals loose all electrical resistance at tempertures below a critical temperature T_c. Many elemental metals are superconductors. Some critical temperatures are: lead T_c = 7.2 K, tin T_c = 3.7 K, niobium T_c = 9.2 K, aluminum T_c = 1.2 K, mercury T_c = 4.2 K, and vanadium T_c = 5.3 K. Compounds can have even higher critical temperatures. $YBa_2Cu_3O_7$ has a critical temperature of T_c = 92 K and $HgBa_2Ca_2Cu_3O_8$ has a critical temperature of

T_c = 133 K. Superconducting wires can carry currents with zero losses up to very high current densities. Above a critical current density J_c, superconductivity is supressed. A typical critical current density is 10^6 A/cm^2. Superconductivity is also supressed by strong magnetic fields.

Superconductor: An element, inter-metallic alloy, or compound that will conduct electricity without resistance below a certain temperature. However, this applies only to direct current (DC) electricity and to finite amounts of current. All known superconductors are solids. None are gases or liquids. And all require extreme cold to enter a superconductive state. Once set in motion, current will flow forever in a closed loop of superconducting material—making it the closest thing to perpetual motion in nature. Scientists refer to superconductivity as a "macroscopic quantum phenomenon". In addition to being classified Type 1 and Type 2, superconductors can be categorized further by their dimensionality. Most are 3-D. But some compounds, like surface-doped $NaWO_3$ and some organic superconductors are 2-D. Li_2CuO_2 and single-walled carbon nano-tubes have shown rare 1-D superconductivity. In addition to repelling magnetic fields, enhanced thermal conductivity, higher optical reflectivity and reduced surface friction are also properties of superconductors. The term "superconductor" is also used in some instances to refer to materials that have near infinite thermal conductivity—such as carbon nanotubes. However, on this website it is used in the context of electrical conductivity only.

An object or substance that conducts electricity with zero resistance. Superconductors, materials that have no resistance to the flow of electricity, are one of the last great frontiers of scientific discovery. Not only have the limits of superconductivity not yet been reached, but the theories that explain superconductor behavior seem to be constantly under review. In 1911 superconductivity was first observed in mercury by Dutch physicist Heike Kamerlingh Onnes of Leiden University (shown above). When he cooled it to the temperature of liquid helium, 4 degrees Kelvin (-452F, -269C), its resistance suddenly disappeared. The Kelvin scale represents an "absolute" scale of temperature. Thus, it was necessary for Onnes to come within 4 degrees of the coldest temperature that is theoretically attainable to witness the phenomenon of superconductivity. Later, in 1913, he won a Nobel Prize in physics for his research in this area. The next great milestone in understanding how matter behaves at extreme cold temperatures occurred in 1933. German researchers Walter Meissner (above) and Robert Ochsenfeld discovered that a superconducting material will repel a magnetic field (below graphic). A magnet moving by a conductor induces currents in the conductor. This is the principle upon which the electric generator operates. But, in a superconductor the induced currents exactly mirror the field that would have otherwise penetrated the superconducting material—causing the magnet to be repulsed. This phenomenon is known as strong diamagnetism and is today often referred to as the "Meissner effect" (an eponym). The Meissner effect is so strong that a magnet can actually be levitated over a superconductive material.

In subsequent decades other superconducting

metals, alloys and compounds were discovered. In 1941 niobium-nitride was found to superconduct at 16 K. In 1953 vanadium-silicon displayed superconductive properties at 17.5 K. And, in 1962 scientists at Westinghouse developed the first commercial superconducting wire, an alloy of niobium and titanium (NbTi). High-energy, particle-accelerator electromagnets made of copper-clad niobium-titanium were then developed in the 1960s at the Rutherford-Appleton Laboratory in the UK, and were first employed in a superconducting accelerator at the Fermilab Tevatron in the US in 1987. The first widely-accepted theoretical understanding of superconductivity was advanced in 1957 by American physicists John Bardeen, Leon Cooper, and John Schrieffer (above). Their *Theories of Superconductivity* became know as the BCS theory—derived from the first letter of each man's last name—and won them a Nobel prize in 1972. The mathematically-complex BCS theory explained superconductivity at temperatures close to absolute zero for elements and simple alloys. However, at higher temperatures and with different superconductor systems, the BCS theory has subsequently become inadequate to fully explain how superconductivity is occurring.

Another significant theoretical advancement came in 1962 when Brian D. Josephson (above), a graduate student at Cambridge University, predicted that electrical current would flow between 2 superconducting materials—even when they are separated by a non-superconductor or insulator. His prediction was later confirmed and won him a share of the 1973 Nobel Prize in Physics. This tunneling phenomenon is today known as the

"Josephson effect" and has been applied to electronic devices such as the SQUID, an instrument capabable of detecting even the weakest magnetic fields. The 1980's were a decade of unrivaled discovery in the field of superconductivity. In 1964 Bill Little of Stanford University had suggested the possibility of organic (carbon-based) superconductors. The first of these theoretical superconductors was successfully synthesized in 1980 by Danish researcher Klaus Bechgaard of the University of Copenhagen and 3 French team members. $(TMTSF)_2PF_6$ had to be cooled to an incredibly cold 1.2K transition temperature (known as Tc) and subjected to high pressure to superconduct. But, its mere existence proved the possibility of "designer" molecules—molecules fashioned to perform in a predictable way.

Then, in 1986, a truly breakthrough discovery was made in the field of superconductivity. Alex Müller and Georg Bednorz (above), researchers at the IBM Research Laboratory in Rüschlikon, Switzerland, created a brittle ceramic compound that superconducted at the highest temperature then known: 30 K. What made this discovery so remarkable was that ceramics are normally insulators. They don't conduct electricity well at all. So, researchers had not considered them as possible high-temperature superconductor candidates. The Lanthanum, Barium, Copper and Oxygen compound that Müller and Bednorz synthesized, behaved in a not-as-yet-understood way. The discovery of this first of the superconducting copper-oxides (cuprates) won the 2 men a Nobel Prize the following year. It was later found that tiny amounts of this material were actually superconducting at 58 K, due to a small amount of lead having been added as a calibration standard—making the discovery even more noteworthy.

Müller and Bednorz' discovery triggered a flurry of activity in the field of superconductivity. Researchers around the world began "cooking" up ceramics of every imaginable combination in a quest for higher and higher Tc's. In January of 1987 a research team at the University of Alabama-Huntsville substituted Yttrium for Lanthanum in the Müller and Bednorz molecule and achieved an incredible 92 K Tc. For the first time a material (today referred to as YBCO) had been found that would superconduct at temperatures warmer than liquid nitrogen—a commonly available coolant. Additional milestones have since been achieved using exotic—and often toxic—elements in the base perovskite ceramic. The current class (or "system") of ceramic superconductors with the highest transition temperatures are the mercuric-cuprates. The first synthesis of one of these compounds was achieved in 1993 by Prof. Dr. Ulker Onbasli at the University of Colorado and by the team of A. Schilling, M. Cantoni, J.D. Guo, and H.R. Ott of Zurich, Switzerland. The world record Tc of 138 K is now held by a thallium-doped, mercuric-cuprate comprised of the elements Mercury, Thallium, Barium, Calcium, Copper and Oxygen. The Tc of this ceramic superconductor was confirmed by Dr. Ron Goldfarb at the National Institute of Standards and Technology-Colorado in February of 1994. Under extreme pressure its Tc can be coaxed up even higher—approximately 25 to 30 degrees more at 300,000 atmospheres. The first company to capitalize on high-temperature superconductors was Illinois

Superconductor (today known as ISCO International), formed in 1989. This amalgam of government, private-industry and academic interests introduced a depth sensor for medical equipment that was able to operate at liquid nitrogen temperatures (~ 77K). While no significant advancements in superconductor Tc's have been achieved in recent years, other discoveries of equal importance have been made. In 1997 researchers discovered that at a temperature very near absolute zero an alloy of gold and indium was both a superconductor and a natural magnet. Conventional wisdom held that a material with such properties could not exist! Since then, over a half-dozen such compounds have been found. Recent years have also seen the discovery of the first high-temperature superconductor that does NOT contain any copper (2000), and the first all-metal perovskite superconductor (2001). Also in 2001 a material that had been sitting on laboratory shelves for decades was found to be an extraordinary new superconductor. Japanese researchers measured the transition temperature of magnesium diboride at 39 Kelvin—far above the highest Tc of any of the elemental or binary alloy superconductors. While 39 K is still well below the Tc's of the "warm" ceramic superconductors, subsequent refinements in the way MgB_2 is fabricated have opened the door to its use in industrial applications. Laboratory testing has found MgB_2 will outperform NbTi and Nb_3Sn wires in high magnetic field applications like MRI. Startling—and serendipitous—discoveries like these are forcing scientists to continually re-examine longstanding theories on superconductivity and to consider heretofore-unimagined combinations of elements.

Magnetic-levitation is an application where superconductors perform extremely well. Transport vehicles such as trains can be made to "float" on strong superconducting magnets, virtually eliminating friction between the train and its tracks. Not only would conventional electromagnets waste much of the electrical energy as heat, they would have to be physically much larger than superconducting magnets. A landmark for the commercial use of MAGLEV technology occurred in 1990 when it gained the status of a nationally-funded project in Japan. The Minister of Transport authorized construction of the Yamanashi Maglev Test Line which opened on April 3, 1997. In December 2003, the MLX01 test vehicle (shown above) attained an incredible speed of 361 mph (581 kph). Although the technology has now been proven, the wider use of MAGLEV vehicles has been constrained by political and environmental concerns (strong magnetic fields can create a bio-hazard). The world's first MAGLEV train to be adopted into commercial service, a shuttle in Birmingham, England, shut down in 1997 after operating for 11 years. A Sino-German maglev is currently operating over a 30 km course at Pudong International Airport in Shanghai, China. The U.S. plans to have its first (non-superconducting) Maglev train in operation this year on a Virginia college campus. Click this link for a website that lists other uses for MAGLEV.

An area where superconductors can perform a life-saving function is in the field of biomagnetism. Doctors need a non-invasive means of determining what's going on inside the human body. By impinging a strong superconductor-derived magnetic field into the body, hydrogen atoms that exist in the body's water and fat molecules are forced to accept energy from the magnetic field. They then release this energy at a frequency that can be detected and displayed graphically by a computer. Magnetic Resonance Imaging (MRI) was actually discovered in the mid 1940's. But, the first MRI exam on a human being was not performed until July 3, 1977. And, it took almost five hours to produce one image! Today's faster computers process the data in much less time. The Korean Superconductivity Group within KRISS has carried biomagnetic technology a step further with the development of a double-relaxation oscillation SQUID (Superconducting quantum Interference Device) for use in Magnetoencephalography. SQUID's are capable of sensing a change in a magnetic field over a billion times weaker than the force that moves the needle on a compass (compass: 5e-5T, SQUID: e-14T.). With this technology, the body can be probed to certain depths without the need for the strong magnetic fields associated with MRI's.

Probably the one event, more than any other, that has been responsible for putting "superconductors" into the American lexicon was the Superconducting Super-Collider project planned for construction in Ellis county, Texas. Though Congress cancelled the multi-billion dollar effort in 1993, the concept of such a large, high-energy collider would never have been viable without superconductors. High-energy particle research hinges on being able to accelerate sub-atomic particles to nearly the speed of light. Superconductor magnets make this possible. CERN, a consortium of several European nations, is doing something similar with its Large Hadron Collider (LHC) now under construction along the Franco-Swiss border. Other related web sites worth visiting include the proton-antiproton collider page at Fermilab. This was the first facility to use superconducting magnets. Get information on the electron-proton collider HERA at the German lab pages of DESY (with English text). Lastly, Brookhaven National Laboratory features a page dedicated to its RHIC heavy-ion collider. Electric generators made with superconducting wire are far more efficient than conventional generators wound with copper wire. In fact, their efficiency is above 99 per cent and their size about half that of conventional generators. These facts make them very lucrative ventures for power utilities. General Electric has estimated the potential worldwide market for superconducting generators in the next decade at around $20-30 billion dollars. Late in 2002 GE Power Systems received $12.3 million in funding from the U.S. Department of Energy to move high-temperature superconducting generator technology toward full commercialization. Other commercial power projects in the works that employ superconductor technology include energy storage to enhance power stability. American Superconductor Corp. received an order from Alliant Energy in late March 2000 to install a Distributed Superconducting Magnetic Energy Storage System (D-SMES) in Wisconsin. Just one of these 6 D-SMES units has a power reserve of over 3 million watts, which can be retrieved whenever there is a need to stabilize line voltage during a disturbance in the power grid. AMSC has also installed more than 22 of its D-VAR systems to provide instantaneous reactive power support.

Recently, power utilities have also begun to

use superconductor-based transformers and "fault limiters". The Swiss-Swedish company ABB was the first to connect a superconducting transformer to a utility power network in March of 1997. ABB also recently announced the development of a 6.4MVA (mega-volt-ampere) fault current limiter—the most powerful in the world. This new generation of HTS superconducting fault limiters is being called upon due to their ability to respond in just thousandths of a second to limit tens of thousands of amperes of current. Advanced Ceramics Limited is another of several companies that makes BSCCO type fault limiters. Intermagnetics General recently completed tests on its largest (15kv class) power-utility-size fault limiter at a Southern California Edison (SCE) substation near Norwalk, California. And, both the US and Japan have plans to replace underground copper power cables with superconducting BSCCO cable-in-conduit cooled with liquid nitrogen. By doing this, more current can be routed through existing cable tunnels. In one instance 250 pounds of superconducting wire replaced 18,000 pounds of vintage copper wire, making it over 7000 per cent more space-efficient.

An idealized application for superconductors is to employ them in the transmission of commercial power to cities. However, due to the high cost and impracticality of cooling miles of superconducting wire to cryogenic temperatures, this has only happened with short "test runs". In May of 2001 some 150,000 residents of Copenhagen, Denmark, began receiving their electricity through HTS (high-temperature superconducting) material. That cable was only 30 meters long, but proved adequate for testing purposes. In the summer of 2001 Pirelli completed installation of three 400-foot HTS cables for Detroit Edison at the Frisbie Substation capable of delivering 100 million watts of power. This marked the first time commercial power has been delivered to customers of a US power utility through superconducting wire. Intermagnetics General has announced that its IGC-SuperPower subsidiary has joined with BOC and Sumitomo Electric in a $26 million project to install an underground, HTS power cable in Albany, New York, in Niagara Mohawk Power Corporation's power grid. The 350-meter cable, believed to be four times the length of any previously installed HTS cable, will be designed to provide more power and operate at significantly lower loss levels than other HTS installations.

The National Science Foundation, along with NASA and DARPA and various universities, are currently researching "petaflop" computers. A petaflop is a thousand-trillion floating point operations per second. Today's fastest computing operations have only reached "teraflop" speeds—trillions of operations per second. Currently the fastest is the IBM Blue Gene/L running at 70.7 teraflops per second (multiple CPU's). The fastest single processor is a Lenslet optical DSP running at 8 teraflops. It has been conjectured that devices on the order of 50 nanometers in size along with unconventional switching mechanisms, such as the Josephson junctions associated with superconductors, will be necessary to achieve such blistering speeds. TRW researchers (now Northrop Grumman) have quantified this further by predicting that 100 billion Josephson junctions on 4000 microprocessors will be necessary to reach 32 petabits per second. These Josephson junctions are incorporated into field-effect transistors which then become part of the logic circuits within the processors. Recently it was demonstrated at the Weizmann Institute in Israel that the tiny magnetic fields that penetrate Type 2 superconductors can be used for storing and retrieving digital information. It is, however, not a foregone conclusion that computers of the future will be built around superconducting devices. Competing technologies, such as quantum (DELTT) transistors, high-density molecule-scale processors, and DNA-based processing also have the potential to achieve petaflop benchmarks. In the electronics industry, ultra-high-performance filters are now being built. Since superconducting wire has near zero resistance, even at high frequencies, many more filter stages can be employed to achive a desired frequency response. This translates into an ability to pass desired frequencies and block undesirable frequencies in high-congestion rf (radio frequency) applications such as cellular telephone systems. ISCO International and Superconductor Technologies are companies currently offering such filters. Superconductors have also found widespread applications in the military. HTSC SQUIDS are being used by the US NAVY to detect mines and submarines. And, significantly smaller motors are being built for NAVY ships using superconducting wire and "tape". In mid-July, 2001, American Superconductor unveiled a 5000-horsepower motor made with superconducting wire (below). The military is also looking at using superconductive tape as a means of reducing the length of very low frequency antennas employed on submarines. Normally, the lower the frequency, the longer an antenna must be. However, inserting a coil of wire ahead of the antenna will make it function as if it were much longer. Unfortunately, this loading coil also increases system losses by adding the resistance in the coil's wire. Using superconductive materials can significantly reduce losses in this coil. The Electronic Materials and Devices Research Group at University of Birmingham (UK) is credited with creating the first superconducting microwave antenna. Applications engineers suggest that superconducting carbon nanotubes might be an ideal nano-antenna for high-gigahertz and terahertz frequencies, once a method of achieving zero "on tube" contact resistance is perfected.

The most ignominious military use of superconductors may come with the deployment of "E-bombs". These are devices that make use of strong, superconductor-derived magnetic fields to create a fast, high-intensity electro-magnetic pulse (EMP) to disable an enemy's electronic equipment. Such a device saw its first use in wartime in March 2003 when US Forces attacked an Iraqi broadcast facility. The future melding of superconductors into our daily lives will also depend to a great degree on advancements in the field of cryogenic cooling. New, high-efficiency magnetocaloric-effect compounds such as gadolinium-silicon-germanium are expected to enter the marketplace soon. Such materials should make possible compact, refrigeration units to facilitate additional HTS applications.

Superelastics: Those materials that can be deformed to eight percent strain. However, it will immediately return to its original shape upon unloading.

Superintelligence: An intellect that is much smarter than the best human brains in practically every field, including scientific creativity, general wisdom and social skills. This definition leaves open how the superintelligence is implemented: it could be a digital computer, an ensemble of networked computers, cultured cortical tissue or what have you. It also leaves open whether the superintelligence is conscious and has subjective experiences.

Superlattice Nanowire Pattern (transfer): [SNAP] a technique for producing "Ultra High Density Nanowire Lattices and Circuits". See Researchers Discover How to Make the Smallest, Most Perfect, Densest Nanowire Lattices-And It's a SNAP

Superlattice: Nanowire interwoven bundles of nanowires using substances with different compositions and properties.

Superlattices: Artificial metallic superlattices are multilayered thin films, prepared by alternately depositing two elements using vacuum deposition or sputtering techniques. A wide spectrum of elements and compounds are suitable for deposition into superlattice structures, and the range of properties displayed by the resulting superstructures is greatly dependent upon the properties of both individual lattices as well as the interaction between them.

Supernova: An explosion of a massive star. One of the most spectacular happenings in the Universe. Nuclear reactions take place in the cores of the stars. These create heavier atomic nuclei from the lighter ones (helium and hydrogen). If the mass of the star is large enough, due to the gravitational force, the star eventually collapses when the pressure provided by the nuclear reactions in its core is not large enough to prevent the gravitational collapse.

This happens when the nuclear fuel for rections depletes and when the heavy elements in the star's core are abdundant. During the collapse, the matter from the star surface bounces off its core and scatters out in the Universe. Thus, heavy elements (atoms) are sent throughout the Universe, some of them finishing at the planets which will once perhaps host life.

Superposition: A quantum mechanical phenomena in which a system exists in more than one state simultaneously.

Surety: A level of confidence that a system will operate exactly as planned under both expected and unexpected circumstances. Surety attributes include reliability in normal environments, safety of people and surroundings in abnormal environments, and security and use control in malevolent environments.

Surface Acoustic Wave (SAW) Sensors: Similar to bulk acoustic wave sensors but operating at a higher frequency of oscillation. The resonators in SAW sensors oscillate usually at 260 MHz.

Surface Micromachining: An additive fabrication technique that involves the building of a device on the top surface of a supportingsubstrate. This technique is relatively independent of substrate; therefore, it can be easily mixed with other fabrication techniques that modify the substrate first.

Surface Passivation: The rendering of electric bonds or contacts ineffective by chemical methods, especially oxidation. It is of critical importance to MEMS reliability.

Surface Plasmon Resonance [microscopy]:. Surface plasmons are optically stimulated by laser illumination of a metal film. These electromagnetic waves travel along the interface between the metal and a dielectric layer. The magnitude of the electromagnetic field close to the surface is extraordinarily sensitive to interface processes—in the BiaCore instrument, to receptor/ ligand interactions, which cause local refractive index changes. Since the vertical resolution of the surface plasmons extends from subnanometer to hundreds of nanometers, surface plasmon microscopy is potentially useful for the study of cell membranes, and transport and trafficking processes involving the membrane, as well as for studies of cell-nanofabricated surface interactions. [National Center for Research Resources "Integrated Genomics Technologies Workshop Report" Jan 1999]

Surface Plasmon: A collective motion of electrons in the surface of a metal conductor, excited by the impact of light of appropriate wavelength at a particular angle.

Surface Roughness: A characteristic of a surface that is found by measuring deviations of the actual surface from its intended shape. The surface roughness of a wafer is the difference between the highest and the deepest surface feature.

Surface Science: A branch of solid state (condensed matter) physics bordering with chemistry. Surface science deals with processes occurring on the surfaces of materials and properties of surfaces. Many techniques of nanoscience and nanotechnology were first created or further developed by surface scientists (atomic force microscope, scanning tunneling microscope, ultra-high vacuum equipment...). Sometimes a term *surface nanotechnology* is used.

Surface Tension: The force acting on a liquid-gas interface resulting in thin film on the surface.

Surface-Mount Technology (SMT): A sophisticated microelectronic packaging technology in which packages are small enough to be soldered using eutectic alloys on the surface of a printed-circuit board.

Surfactant: Short for "SURFace ACTive AgeNT". A surfactant is a molecule with an amphiphilic character that has a tendency to accumulate at interfaces. Surfactants is a larger class of molecules which have a significant technological and biological importance. Generally these molecules consist of a hydrophilic (water soluble) and a hydrophobic (water insoluble) part. This amphiphilic nature of surfactants is responsible for their association behaviour in solution (micelles, bilayers, vesicles, etc.) and their accumulation at interfaces (gas/liquid, liquid/liquid or liquid/solid). The hydrophobic part usually consists of hydrocarbon or fluorocarbon chains, while the hydrophilic part consists of a polar group (-OH, -COOH, $-NH_3+$, $-PO_4-(CH_2)_2NH_3+$ etc.). The association behaviour of surfactants in solution and their affinity for interfaces is determined by the physical and chemical properties of the hydrophobic and hydrophilic groups, respectively. The size and shape of the hydrocarbon moiety and the size, charge and hydration of the hydrophilic head group are of utmost importance in this respect. Depending on the balance between these properties a wide variety of self-assembled structures, both at interfaces and in bulk, have been observed. The driving force behind the association is the reduction of the free energy of the system. Therefore, when a surfactant comes in contact with water it accumulates at the air/water interface causing a decrease in the surface tension of water. Surfactants are the active ingredients in soap, since the hydrophobic component sticks to grease and dirt while the hydrophilic section sticks to the water.

Susceptibility: A measure of the relative amount of induced magnetism in a material. Magnetic susceptibility is often used in lieu of resistance measurements to determine the transition temperature of a superconductor. Although, on occasion, the two techniques produce very different Tc's. In a typical superconductor, the (arbitrary) value of susceptibility will change from zero to a negative number as the temperature drops through Tc. However, in some materials it changes from positive to negative, as paramagnetism yields to diamagnetism.

Susceptor: A platen on which semiconductor wafers are heated during thermal processes such as epitaxy and chemical-vapor deposition (CVD).

Sustaining Technology: A technology associated with a developed market. It is the current stable technology product platform in an industry or firm setting.In well-managed companies that supply this technology, the products advance as rapidly as improvements in science and engineering allow. Managers carefully determine the desires of their customers and plan new engineering projects to satisfy those needs quickly and cost-effectively.

Swarms: Active materials can theoretically be made entirely of machines. These are sometimes called swarms since they consist of large numbers of identical simple machines that grasp and release each other and exchange power and information to achieve complex goals. Swarms change shape and exert force on their environment under software control. Although some physical prototypes have been built, at least one patent issued, and many simulations run, swarm potential capabilities are not well analyzed or understood. We briefly discuss some concepts here. Michael proposes brick-shaped machines of various sizes that slide past each other to assume a variety of shapes. He has generated a large number of videos showing computer simulations of simple motions. Although his web site contains rather extravagant claims, this work has received a U.K. patent. Yim built a small swarm with macroscopic (size in inches) components called polypod, built a simulator of polypod, and programmed it to move in various ways to study locomotion. There are two brick shaped components in polypod, one of which has two prismatic joints linked by a revolute joint. The second component is a cubic connector with no mechanical motion. Polypod is programmed by tables for each member of the swarm. Each member is programmed to move at various speeds in each degree of freedom for certain amounts of time. The swarm components are implicitly synchronized so there is no clock signal. Hall proposes a swarm with 10 micron dodecahedral components each with 12 arms that can move in and out, rotate a little, and grab and release each other. This concept is called the "utility fog." Hall estimates that the utility fog would have a density of 0.2, tensile strength of 1000 psi in action and 100,000 psi in a passive mode, and have a maximum shear rate of 100 km/second/meter. Bishop proposes a swarm consisting of 100 nanometer brick-shaped components that slide past each other to change shape. Globus proposes a swarm with two kinds of components — edges and nodes. The terms "node" and "edge" are chosen to correspond to those in graph theory. The roughly spherical nodes are capable of attaching to five edges (for a tetrahedral geometry with one free edge per node) and rotating each edge in pitch and yaw. The rod-like edges are capable of changing length, rotating around their long axis, and attaching/detaching to/from nodes.

S-Wave: A form of electron pairing in which the electrons travel together in spherical orbits, but in opposite directions.

Swedish Viking: A scientific satellite launched in February 1986. Viking carried the Canadian Space Agency's Ultraviolet Auroral Imager which provided important new information about the aurora borealis.

Synapse: A structure that transmits signals from a neuron to an adjacent neuron (or other cell).

Synchrotron: A large device that imparts extremely high velocity to charged particles by combining a high-frequency electric field and a low-frequency magnetic field. Used as a source for X-rays.

Synchrotrons: Devices for accelerating protons or electrons in closed orbits where the accelerating voltage and magnetic field strength varies (the accelerating voltage is held constant for electrons) in order to keep the orbit radius constant.

Synthesis: The CAD operation of automatically generating an IC layout from a high-level description (e.g. using a hardware description language) of the desired circuit function and behavior. The production of a specific *molecular* structure by a series of chemical *reactions.*

Synthespian: An artificial actor, for example a 3D model animated by motion capture from a real actor or a computer program.

Synthetic Aperture Radar (SAR): An instrument that uses radar to see through dark, clouds and fog. This type of equipment is on board only three satellites in the world, including Radarsat and Lacrosse.

System: In scientific usage, usually equivalent to "a collection of matter and energy being analyzed as a unit." In engineering usage, usually equivalent to "a set of components working together to serve a set of purposes."

Systematic Error: An error that always occurs in the same direction. A component of the error of measurement, which, in the course of a number of measurements of the same measurand, remains constant or varies in a predictable way. Systematic errors and their causes may be known or unknown.

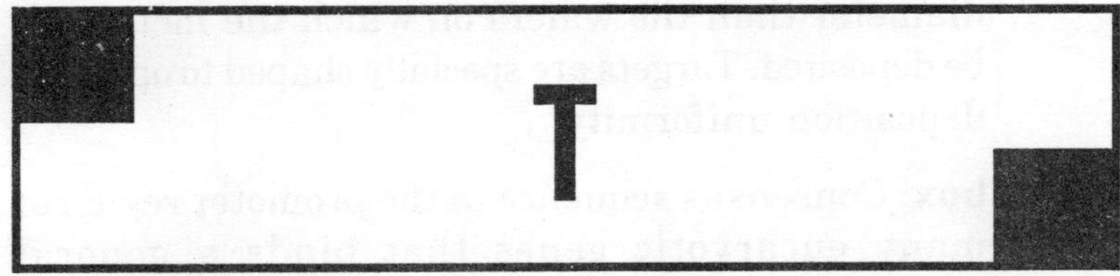

T Lymphocyte (T cell): Type of lymphocyte responsible for cell-mediated immunity; includes both cytotoxic T cells and helper T cells.

TAB Bonding: Tape automated bonding; semiconductor packaging technique that uses a tiny lead-frame to connect circuitry on the surface of the chip to a substrate instead of wire bonds.

Tape-Automated Bonding (TAB): A bonding technique using a transparent flexible polyester tape that has tracks on which unpackaged ICs are attached and stored in a reel. Silver epoxy is screen-printed onto the pads of the tape to which the device's leads are to be connected. The tape is fed through an automatic machine, which pushes the device, and the TAB leads into the epoxy. On curing, the silver-loaded epoxy forms electrical connections between the TAB leads and the pads on the substrate.

Target Validation Technologies: A number of technologies including downregulation of gene expression (gene knockdown, antisense, ribozymes and zinc finger proteins), protein inhibition (phage libraries and antibodies), cellular assays, chemical genetics, and combinatorial biology are linked with target validation. The integration of various technologies is another challenge Targets glossary

Target: The metal source for a sputtering process. Targets are mostly round, thick metal pieces larger in diameter than the wafers on which the metal will be deposited. Targets are specially shaped to optimize deposition uniformity.

TATA box: Consensus sequence in the promoter region of many eucaryotic genes that binds a general transcription factor and hence specifies the position where transcription is initiated.

Taxonomy: A general method for classification. An example of a MEMS taxonomy is that proposed by P. McWhorter, which classifies MEMS applications according to the following categories: (1) non-moving components, (2) moving components that do not touch each other, (3) components that move and impact each other, and (4) components that move and rub against each other.

Tc: The scientific notation representing the critical transition temperature below which a material begins to superconduct. The sudden loss of resistance in a superconductive medium may occur across a range as small as 20 millionths of a degree or, in the case of some stoichiometrically imperfect compounds, tens of degrees.

TCE (Thermal-Coefficient-of Expansion): Defined as the rate at which a material expands and contracts as a temperature rises or falls. It is measured as change in length per change in temperature. The TCE of adjacent thin films in semiconductor microfabrication have to be similar to avoid failure-causing stresses induced at room temperature.

Technocyte: A nanoscale artificial device (especially a nanite) in the human bloodstream used for repairs, cancer protection, as an artificial immune system or for other uses.

Technofobics: Those who have a phobia to technology, and/or to advances in technology.

Technology: The application of knowledge and understanding in order to control processes and fabricate products, from the Greek root *techni*, meaning 'art'.

Technology File: A software file used to design the fabrication process steps necessary to achieve a MEMS or IC component.

Technology Node: Also called minimum feature sizes of a device. It refers to the line width of a transistor and gives an insight into the sophistication of a semiconductor fabrication process. The technology complexity of an IC product is represented by a technology node E.g. a 0.13-μm (130-nm), 0.15-μm (150-nm), etc., technology node. The smaller the technology node the more the number of transistors can be put on a single die. At present, most MEMS devices are in the 1-μm technology node while IC processing technology has reached 0.09 μm.

Technology Push: Often referred to as the domain of a market visionary. Technology push occurs when a firm introduces a new product that they think will be beneficial and of interest to a consumer before the consumer is even aware of it. Technology push is more risky than market pull. The producer has to educate the consumer of the product's benefits.

Technology Validation: This Center offers technology validation services to a broad range of Commercial Investigators, from emerging Biotechs to large Pharma and Instrument Manufacturers. In the TVC, technology validation is defined as confirmation of the appropriateness, effectiveness, robustness, accuracy, and reproducibility of a particular technology when applied to a specific task. Successful validation is defined by technical and statistical parameters established in advance by INCAPS in consultation with the customer.

Technology/Product Paradigm: The technology a company utilizes to form a "core product" that acts as a platform from which many application-specific products can be developed. It is a matrix describing the relationship between the various technologies and products of a company.

Technology: The application of knowledge and understanding in order to control processes and fabricate products, from the Greek root *techni*, meaning 'art'.

Telecalibration: The use of information technology to offer remote calibration services for metrology equipment using a communication network.

Telescope: A device used in astronomy to see distant objects. Most telescopes use lenses and mirrors to magnify light coming from phenomena deep in space. This makes the objects look bigger and closer. Newer telescopes, however, are using radio waves, infrared light, laser, and radar technologies.

Telomere: End of a chromosome, associated with a characteristic DNA sequence that is replicated in a special way. Counteracts the tendency of the chromosome otherwise to shorten with each round of replication. (From Greek *telos*, end.)

Telophase: Final stage of mitosis in which the two sets of separated chromosomes decondense and become enclosed by nuclear envelopes.

Temperature Coefficient of Resistance (TCR): The rate of change of resistance due to temperature variation. Mathematically, TCR = (R/R)/ T, where refers to a change.

Temperature Sensitive (ts) Mutant: Organism or cell carrying a genetically altered protein (or RNA molecule) that performs normally at one temperature but is abnormal at another (usually higher) temperature.

Temperature: A system in which internal vibrational modes have equilibrated with one another can be said to have a particular temperature. Two systems A and B are said to be at different temperatures if, when brought into contact, *heat* flows from (say) A to B, increasing the *thermal energy* of B at the expense of the thermal energy of A.

Template: An external source of information or structure often used in manufacturing and mass production. As a nanoscale example, biological systems are known to use nucleic acids as a templates for protein synthesis.

Templating: The directed assembly of materials through interaction with a template.

Tensile Strain: An increase in a dimension (length, width, etc.) of a body when it is subjected to an external tensile (stretching) force.

Tensile Strength (TS): The maximum engineering stress, in tension, sustainable without fracture; also called"ultimate (tensile) strength".

Tensile: Pertaining to the stress (i.e., tension) resulting from the elongation of an elastic body.

Tensiometer: Force measuring instrument for determining surface / interfacial tensions of liquids, and dynamic contact angles of (regular shaped) solids, fibers, powders and porous materials. The measurement of surface / interfacial tension of liquids and contact angles of liquids against solids as performed by a tensiometer is based on force measurements of the interaction of a solid probe with the interface of a liquid or between 2 immiscible liquids. In these experiments a probe is hung on a balance and brought into contact with the liquid interface. The forces experienced by the balance as the probe interacts with the surface of the liquid can be used to calculate surface / interfacial tension or contact angles. The forces present in this situation depend on the following factors; size and shape of the probe, contact angle of the liquid/solid interaction and surface tension of the liquid. The size and shape of the probe are easily controlled. For surface / interfacial tension measurements the contact angle of the liquid towards the probe is controlled to be zero (complete wetting). This is achieved by using probes with high energy surfaces. For example, probes made of a platinum/iridium alloy insure complete wetting and they are very easily cleaned in a reliable way. The mathematical interpretation of the force measurements depends on the shape of the probe used. Two types of probes are commonly used for surface / interfacial tension measurements, the Wilhelmy plate and DuNouy Ring. The Wilhelmy plate method utilizes the interaction of a platinum plate with the liquid interface being tested. The calculations for this technique are based on the geometry of a fully wetted plate in contact with, but not submerged in, the liquid. In this method the position of the probe relative to the surface is significant. Therefore the measurement is made at the so called zero depth of immersion of the probe, and the force acting on the probe at this position is registered and can directly be used to calculate the surface tension of the liquid when the perimeter of the plate is accurately known. If only limited quantity of the liquid to be tested is available one may consider using a thin round platinum rod as the probe. In such a case the measurement is exactly the same as with the Wilhelmy plate, but the probe dimensions of the probe will be smaller which affects the accuracy of the measurement and hence also might affect the reproducibility of the end results. The Du Nouy ring method utilizes the interaction of a platinum ring with the liquid interface being tested. The ring is submerged below the interface and subsequently raised upwards. As the ring moves upwards it raises a meniscus of the liquid. Eventually this meniscus tears from the ring and returns to it's original position. Prior to this event, the volume, and thus the force exerted, of the meniscus passes through a maximum value and begins to diminish prior to the actually tearing event. The calculation of surface or interfacial tension by this technique is based on the measurement of this maximum force. As an additional volume of liquid which is raised due to the proximity of one side of the ring to the other mathematical corrections are needed in order to obtain the correct surface / interfacial tension values. he tensiometric method for measuring dynamic contact angles (Advancing and Receding contact angles) measures the forces that are present when a sample of solid is brought into contact with a test liquid. If the forces of interaction, geometry of the solid and surface tension of the liquid are known the contact angle may be calculated. The solid sample has to have a symmetrical and regular shape (rod, cube, round rod, rectangle, wire etc.) in order to use this method. The advantages of this method are that the actual measurement is very simple to perform, one measurement gives an average for the whole sample and that wetting characteristics of single fibers can easily be obtained. he Tensiometer instrument also enables to determine the wettability characteristics i.e. contact angles of powders and porous solids by using the Washburn method (Powder Wettability Measurement).

TEOS (TetraEthylOrthoSilicate): A chemical used in CVD deposition of SiO_2. It is commonly used in the industry since the film achieved is conformal.

Terraform: To change the properties of a planet to make it more earthlike, making it possible for humans or other terrestrial organisms to live unaided on it, for example by changing atmospheric composition, pressure, temperature or the climate and introducing a self-sustaining ecosystem. This will most probably be a very long-term project, probably requiring self-replicating technology and megascale engineering. So far Venus and especially Mars looks as the most promising candidates for terraforming in the solar system. [Jack Williamson 1938] [AS] Speculation exists that with the advent of mature MNT that we should be able to accomplish Terraforming a planet such as Mars in years, rather then decades [editor]

Tertiary Structure: Complex three-dimensional form of a macromolecule, especially a protein.

Tesla [T]: Unit of magnetic induction: 1T = 1 weber/m^2 (also, 1T = 10^4 gauss).

Tetrahymena: Genus of ciliated protozoa used in studies of ciliary axonemes, self-splicing RNA, and telomere reproduction.

Tetra-Methyl-Ammonium-Hydroxide (TMAH): A chemical used to develop photoresist after the photo resist-coated wafer is exposed to light through a mask.

The Atom: Atoms are the building blocks of matter. All things are made of arrangements of atoms. There are 118 (with some very short lived ones) different elements in the periodic table, which means that we know of 118 different kinds of atoms.

The Fermi Paradox: "If there are other intelligent beings in the Universe, why aren't they here?" Since it appears to be quite possible for a technological species to spread across the galaxy in less than 10 million years (using von Neumann machines) or otherwise change things on such a large scale that it would be very visible (see Kardaschev types), the lack of such evidence is puzzling or implies that other technological civilizations doen't exist. There have been many attempts to explain this, for example the "Wildlife Preserve" idea (the aliens doesn't want to interfere with younger civilizations), that they transcend and become incomprehensible, that they hide or that they are actually here, hidden on the nanoscale, but the problem with these attempts is that most of them just explain why some aliens would not be apparent.

Thermal Actuator: An electro-thermally actuated MEMS device. A current passed though the device causes

Joule (I^2R) heating resulting in differential thermal expansion of the device, which can be used to extract a relatively large mechanical deformation.

Thermal Blanketing: Like insulation in a house, the covering on a satellite that regulates temperature. A material commonly used for thermal blanketing on a satellite is Mylar.

Thermal Budget: The amount of thermal energy transmitted to a wafer. It is proportional to the product of temperature and time of the heating process.

Thermal Compression Bonding: A method of direct bonding that does not use an intermediary metal or melting, but rather the plastic flow of materials resulting form the combination of heat and pressure. It's used to bond beam-lead devices to packaging substrates.

Thermal Conductivity (ê): For steady-state heat flow, the proportionality constant between the heat flux and the temperature gradient. Also, a parameter characterizing the ability of a material to conducting heat. A measure of the effectiveness of a material as a thermal insulator. A substance with a large thermal conductivity value is a good conductor of heat; one with a small thermal conductivity value is a poor heat conductor, i.e. a good insulator. Its units are W/mK.

Thermal Diffusion: A phenomenon in which a temperature gradient in a mixture of fluids gives rise to a flow of one constituent relative to the mixture as a whole. Also known as thermodiffusion.

Thermal Energy: The *internal energy* present in a system as a result of the energy of thermally equilibrated vibrational modes and other motions (including both kinetic energy and molecular potential energy). The mean thermal energy of a classical harmonic oscillator is kT.

Thermal Expansion Coefficient: The rate of change of length with respect to temperature for a particular material.

Thermal Fatigue: A type of fatigue failure that introduces the cyclic stresses by fluctuating thermal stresses.

Thermal Fluctuations: The *thermal energy* of a system (or of a particular part or mode of motion in a system) has a mean value determined by the *temperature* and by the structure of the system. Statistical deviations about that mean are termed thermal fluctuations; these are of great importance in determining both rates of chemical *reactions* and error rates in nanomechanical systems.

Thermal Materials: The record-setting anisotropic thermal conductivity of buckytubes is enabling applications where heat needs to move from one place to another. Such an application is electronics, particularly advanced computing, where uncooled chips now routinely reach over 100°C. CNI's technology for creating aligned structures and ribbons of buckytubes is a step toward realising incredibly efficient heat conduits. In addition, composites with buckytubes have been shown to dramatically increase the bulk thermal conductivity at small loadings.

Thermal Noise: the vibration and motion of atoms and molecules caused by the fact that they have a temperature above absolute zero. [RCM] Once used as an argument on why MNT could not work. Since refuted: See Thatís impossible! How good scientists reach bad conclusions

Thermal Protection: Thermal protection is crucial for atmospheric reentry and other tasks. The carbon nanotubes under investigation at NASA Ames and elsewhere may play a significant role. Most production processes for carbon nanotubes create a tangled mat of nanotubes that has a very low mass-to-volume ratio. Like graphite, the tubes should withstand high temperatures but the tangled mat should prevent them from ablating. This may lead to high temperature applications.

Thermal Shock: The fracture of a brittle material resulting from stresses introduced by a rapid temperature change.

Thermal Stability: The high thermal stability of cerium oxide results in Ce_2O_3 nanoparticles re-oxidising and remaining active after enhancing the initial combustion cycle. In a diesel engine a significant amount of soot is formed and this may adhere to the internal combustion chamber surfaces as a deposit together with the lubricating oil mist. These deposits in turn affect the operation of the engine by introducing friction losses and variations in the internal surface heat loss behaviour. Although the piston head and chamber walls are comparatively cool in respect to the combustion flame the temperature still oscillates between 200°C and 500°C. This is exactly the range in which the nanoparticle cerium oxide carbon activation temperature lies and hence carbon deposits are gradually oxidised and removed from the engine allowing more efficient operation and reduced fuel consumption.

Thermal Stress: A residual stress introduced inside a body due to a change in ambient temperature.

Thermal Vacuum Chamber: The closest thing to space on Earth. At the David Florida Laboratory, satellites are tested in this chamber which can simulate the temperature and vacuum of space to measure the cool-down and warm-up characteristics of a piece of hardware. This allows engineers to predict how the satellite's hardware might perform under the harsh conditions of space.

Thermistor: A thermally sensitive resistor. It's manufactured using barium-titanate or the oxides of transition metals—manganese, cobalt, copper and nickel. Thermistors are used in measurement and control and temperature compensation.

Thermo-Compression (TC) Bonding: An IC bonding process between a gold wire and a gold metallized substrate using a wedge-shaped tool under accurate and precise combination of heat, force and mechanical vibration.

Thermocouple: A device for measuring temperature. It consists of a pair of wires of electrochemically different metals like copper and iron. The free ends of the wires are connected to a voltmeter, which measures the difference in potential created at the junction of the two metals. A temperature-measuring device, which contains a pair of end-joined dissimilar conductors in which an electromotive force is developed by thermoelectric effects when the joined ends and the free ends of the conductors are a different temperature.

Thermocycler: An instrument that repeatedly cycles through various temperatures required for an iterative, temperature-dependant chemical process such as the polymerase chain reaction.

Thermodynamical Stability: Stability of a certain system with respect to changes in thermodynamical parameters. For nanoscopic systems, the most important parameter is temperature , i.e. stability of a system with respect to increase of temperature. Some of nanoscopic structures are stable only at sufficiently low temperatures, while at higher temperatures their structure tends to degrade. At higher temperatures, the atoms of which a nanostructure is built vibrate with larger amplitudes which may result in degradation of the structure. Additionally, nanostructured material, or a nanoparticle may be unstable when exposed to chemical influences from the surrounding, i.e. atoms and molecules. Some nanostructures are stable only in ultra-high vacuum conditions, which very much limits their use. Thermodynamical and chemical stability is a major concern in many applications involving nanoparticles and nanostructured materials.

Thermodynamics: A field of study embracing energy conversion among various forms, including *heat*, *work*, and *potential* and kinetic energy.

Thermoelastic: Both *stress* and *temperature* changes alter the dimensions of an object having a finite *stiffness* and a nonzero thermal expansion coefficient. Applying a stress then produces a temperature change; this can result in a *heat* flow which then changes the stress: these are thermoelastic effects, and result in losses of *free energy*.

Thermogenic Limit: In medical nanorobotics, the maximum amount of waste heat that may safely be released by a population of in vivo medical nanorobots that are operating within a given tissue volume.

Thermoplastic Polymer: A substance that when molded to a certain shape under appropriate conditions can later be remelted.

Thermoplastic: Plastics that soften or fuse when heated and harden again when cooled.

Thermoset Polymer: A substance that when molded to a certain shape under pressure and high temperatures cannot be softened again or dissolved.

Thermosonic Bonding: Similar to Thermo-compression bonding except for the addition of high-frequency ultrasonic vibration to the tip during bonding. The interface temperature is around 100-150°C and thus avoids the high-temperature effects of thermo-compression bonding that needs a temperature >300°C.

Thick-Film Process: (1) A hybrid microelectronic process where conductors, insulators, and passive components are screened from special pastes onto a substrate. This process is less expensive than the thin-film process that uses photolithographic techniques to define conductors, etc. (2) The use of deposited films of conductive or insulation material, which may be patterned via photolithographic techniques to form electronic components and conductors on a substrate, or used as insulation material between successive layers of components.

Thin Film (Deposition): A method of fabricating ceramic superconductors to more precisely control the growth of the crystalline structure to eliminate grain boundaries and achieve a desired Tc. This can involve Pulsed-Laser Deposition (PLD) or Pulsed-Electron Deposition (PED) of the material. A variation of this technique can be used to increase the Tc of a superconductor by growing it on a supporting material with a smaller interatomic spacing. The supporting material acts as a molecular "girdle" to compress the atomic lattice of the superconductor, thereby raising its transition temperature. Superconductive tape is made using thin film deposition technology.

Thin-Film Transistor (TFT): A transistor whose active, current-carrying layer is a thin film of silicon that is on top of a glass substrate. TFTs are used in flat-panel display since light is able to pass through a transparent glass to be visible to a viewer. They're cheaper than conventional transistors since they don't require lot of clean room time and use inexpensive equipment for manufacture.

Thioester Bond: High-energy bond formed by a condensation reaction between an acid (acyl) group and a thiol group (- SH); seen, for example, in acetyl CoA and in many enzyme-substrate complexes.

Thiol: An SH group, or a molecule containing one. Also known as a sulfhydryl or mercapto group.

Three-axis Stabilized: Done with an internal gyroscope and thrusters. A gyroscope's stable spin can be used as a sensor to tell the satellite when its attitude is changing. The satellite can then correct the problem using thrusters. This is one form of attitude control.

Three-D Geometry Simulator: 3D geometric model of a MEMS device built by simulating the manufacturing process using a mask set and a process-definition file. The resulting solid model is in a standard commercial file format that can be used for visualization, analysis, dynamic modeling, rapid prototyping, etc.—e.g. SUMMiT 3D Modeler from Sandia National Labs (USA).

Three-D: Three-dimensional. The visual display that exhibits width, height, and thickness or depth. Standard 2D computer images and television displays create a flat image with only height and width.

Threshold: The smallest input signal that will cause a readable change in the output signal.

Thrombin: The enzyme that converts fibrinogen to fibrin to form a blood clot. Thrombin circulates in the inactive form, prothrombin.

Thrombogenesis: Blood coagulation involves a complex series of reactions in which various proteins are enzymatically activated in a sequential manner, transforming liquid blood into a gel-like clot which is then stabilized to form a thrombus (clot) consisting of platelets, fibrin, and red cells. The series of reactions is classically divided into two pathways—extrinsic and intrinsic —involving more than a dozen factors that converge on a single common final pathway, resulting in clot formation. Since, these factors are carried by the blood, the vasculoid can regulate their local concentration and thus, prevent thrombogenic pathways from proceeding to completion. Additionally, platelets must undergo adhesion and activation for coagulation to occur. The adhesion of platelets to exposed collagen in injured blood vessels is mediated by a bridging molecule called von Willebrand's factor that is secreted by endothelial cells into plasma, which prevents platelets from detaching under the high shearing stresses developed near vessel walls. The activation of normally quiescent platelets is a complex phenomenon that includes changes in cell shape, increased movement, release of the contents of their granules (containing nucleotidyl phosphates, serotonin, various factors, enzymes and plasma proteins), and aggregation. The most potent activator of platelets in vivo is thrombin, which interacts with a receptor on the platelet plasma membrane, followed by transmembrane signaling and subsequent activation of the cell. Collagen is the other most important platelet activator; ADP can stimulate aggregation but not granule release. In principle, the blood-contacting surfaces of a nanoorgan, or of nanorobots in sufficient bloodstream numbers and concentrations, could activate platelets (and thus either of the two coagulation pathways), but careful choices of materials and of allowable mechanical motions should reduce or eliminate inherent nanodevice thrombogenicity. For example, DLC diamond-coated stents, heart valves and other blood-contacting LVAD surfaces or substrates generally show reduced thrombogenicity and no platelet activation. Sapphire (alumina ceramic) has low thrombogenicity and both platelet adhesion and activation are low. Hemolysis is near-zero for diamond and alumina powders. Additionally, in the vasculoid plateletogenesis and release of platelets could be actively regulated and reduced to minimal levels. Future experiments must determine if ordinary diamondoid surfaces will have to be supplemented with additional antithrombogenic coatings in order to achieve vasculoid performance objectives. If such coatings are required, one simple possibility is surface-immobilized heparin, a ~15 kD straight-chain anionic (acidic) mucopolysaccharide (glycosaminoglycan) that forms polymers of various lengths. Heparin, first discovered in 1916, is produced naturally by human liver mast cells and basophil leukocytes, and inhibits coagulation primarily by accelerating the interaction between antithrombin and thrombin. Nanorobot exteriors can be "heparinized", and thereby rendered thromboresistant by immobilized heparin on all blood-contacting surfaces at ~monolayer surface concentration (e.g., 7-10 pmol/cm^2). Cellulose membranes coated with 3.6 pmol/cm^2 of endothelial-cell-surface heparan sulphate show complete inhibition of platelet adhesion. If satisfactory passive nonthrombogenic surfaces cannot be found, nanorobots might employ any of several active strategies to prevent iatrogenic coagulation.

Through-Hole Components: Components with lead that need insertion through holes in the PCB and are soldered on the opposite side of the PCB. They're more labor-intensive than surface-mount technology (SMT) parts and are losing popularity.

Throughput Time: The ratio of the amount of time taken to complete a production run and the number of acceptable work pieces produced. If it takes 1000 minutes to make 100 pieces then throughput time is 10 minutes.

Throughput: The number of wafers per hour through a semiconductor equipment tool, a production stage, or the factory, assuming 100% equipment uptime at its maximum capacity. It is adjusted for any idle time under certain circumstances.

Through-WaferVias: An interconnect technology used to make 3D MEMS structures. It solves the problem of wafer front-side space shortage and enables high-density packaging.

Thruster: A way of controlling a satellite's attitude. Thrusters usually contain compressed gas that when sent out of the end of the thruster will move the satellite in space. The force of the compressed gas (the action) causes the satellite to move in the opposite direction (the reaction). This is Newton's third law of motion—for every action there is an equal and opposite reaction.

Thylakoid: Flattened sac of membrane in a chloroplast that contains pigment and carries out the light-gathering reactions of photosynthesis. Stacks of thylakoids form the grana of chloroplasts.

Thyristor: The thyristor (also called a silicon-controlled rectifier or SCR) is a solid-state semiconductor device similar to a diode, with an extra terminal which is used to turn it on. Once turned on, the thyristor will remain on (conducting) as long as there is a significant current flowing through it. If the current falls to zero, the device switches off.

The thyristor is a four-layer semiconducting device, with each layer consisting of an alternately N or P-type material, for example N-P-N-P. The main terminals, labelled anode and cathode, are across the full four layers, and the control terminal, called the gate, is attached to one of the middle layers. The operation of a thyristor can be understood in terms of a pair of tightly coupled transistors, arranged to cause the self-latching action. Thyristors are mainly used where high currents and voltages are involved, and are often used to control alternating currents, where the change of sign of the current causes the device to automatically switch off. This is known as *synchronous operation* or Zero Cross operation.

This principle is used to control the desired loading by adjusting the frequency of the sinusoidal input. The range of frequencies is great because there is no limit to the number of cycles a thyristor can perform, and exhibits no "wear out" modes. This is a frequency domain method of control. With phase angle control a thyristor is turned on at a specific and adjustable portion of the cycle of the controlling sinusoidal input. Moving the point at which the thyristor is turned on regulates power output. An example of this method of control is a dimmer switch for lights. The turn on point of a thyristor is controlled to occur at a particular point on the sine curve of the AC supply. The thyristor stays on for the remainder of that cycle and the longer the thyristor stays on, the brighter the light. Fine resolution of output is possible with this method and is suitable for fast responding loads such as tungsten filament lamps or temperature variable resistance loads. Phase-angle control is also essential for inductive loads. The drawback of a thyristor is that, like a diode, it only conducts in one direction. A similar self-latching 5-layer device, called a triac, is able to work in both directions. An earlier gas filled tube device called a Thyratron provided a similar electronic switching capability, where a small control voltage could switch a large current. Modern thyristors can switch large amounts of power (up to megawatts). In the realm of very high power applications, they are still the primary choice. However, in low and medium power (from few tens of watts to few tens of killowatts) they have almost been replaced by other devices with superior switching characteristics like MOSFETs or IGBTs. One major problem of thyristor is that is it not a fully controllable switch in the sense that triggering current direction need to be reversed to switch it off. GTO (Gate Turn-off Thyristor) is another related device which addresses this problem. In high-frequency applications, thyristors are poor candidates due to large switching times arising out of bipolar conduction. MOSFETs, on the other hand, has much faster switching capability because of its unipolar conduction (only majority carriers carry the current).

Tight Junction: Cell-cell junction that seals adjacent epithelial cells together, preventing the passage of most dissolved molecules from one side of the epithelial sheet to the other.

Tight-receptor Structures: A receptor structure in which a bound *ligand* of a particular kind is confined on all sides by repulsive interactions (note that favorable binding energies are compatible with repulsive forces). A tight-receptor structure discriminates strongly against all molecules larger than the target.

Time Constant: The time it takes for the output change to reach 63% of its final value.

$TiSi_2$ (Titatium Silicide): A compound of titanium and silicon that is coated by CVD. It is used to reduce the resistance of the metal contact on the silicon.

Tissue Engineered Medical Products Standards (TEMPS): The standards process for tissue engineered medical products is underway within the American Society for Testing and Materials (ASTM) Committee F-04, Division IV Tissue Engineered Medical Products or Committee F04 on Medical and Surgical Materials and Devices.

The Executive Roster was established. Information describing the process and progress will be presented at this FDA site. Up dated information and current draft documents can be viewed. Additional information may be obtained at the Pittsburgh Tissue Engineering Initiative site. [FDA, Center for Devices & Radiologic Health, US, Tissue Engineered Medical Products Standards TEMPS, 2002]

Tissue Engineering: Generating tissue *in vitro* for clinical applications, such as replacing wounded tissues or impaired organs. A *cell culture* facility is required for cell harvest and expansion. The term "tissue engineering" was coined at an NSF [National Science Foundation] -sponsored meeting in 1987. At a later NSF- sponsored workshop, tissue engineering was defined as "...the application of principles and methods of engineering and life sciences toward fundamental understanding ...and development of biological substitutes to restore, maintain and improve [human] tissue functions." This definition is intended to include procedures where the biological substitutes are cells or combinations of different cells that may be implanted on a scaffold such as natural collagen or as synthetic, biocompatible polymers to form a tissue.

Tissue Engineering: Generating tissue *in vitro* for clinical applications, such as replacing wounded tissues or impaired organs. A cell culture facility is required for cell harvest and expansion. The term "tissue engineering" was coined at an NSF [National Science Foundation] -sponsored meeting in 1987. At a later NSF- sponsored workshop, tissue engineering was defined as "...the application of principles and methods of engineering and life sciences toward fundamental understanding ...and development of biological substitutes to restore, maintain and improve [human] tissue functions." This definition is intended to include procedures where the biological substitutes are cells or combinations of different cells that may be implanted on a scaffold such as natural collagen or as synthetic, biocompatible polymers to form a tissue. ["Tissue Engineering" National Science Foundation].

Tissue Models: Cells, tissues, and organs function in a 3-D environment. Ideally, cell-based models that capture both the 3-D organization and multicellular complexity of the target system provide the most powerful tools for screening the effects of therapeutic candidates. This meeting will bring together tissue engineers who are constructing the 3-D models with biologists who are studying healthy vs. diseased states and pharmacologists who are developing high throughput screening assays. As with 3-D biological models, each specialty provides insight into the complete system.

Titanium Nanopositioner Stages: If the positioning accuracy must be maintained over a prolonged time and temperatures cannot be maintained accurately, use of alternative materials may provide the necessary positioning stability. Titanium's coefficient of thermal expansion (0.9×10^{-5} mm/mm per °C) represents a significant improvement over aluminum alloys and offers the additional benefit of higher strength –improving nanopositioner performance.

Tolerance: The total permissible variation of a quantity from a designated value.

Top Down Molding: [AKA: mechanical nanotechnology] Carving and fabricating small materials and components by using larger objects such as our hands, tools and lasers, respectively. [NTN] Opposite of Bottom Up.

Top Down Nanosystems: These systems utilize metaphors such as "chip" (large scale integrated nanosystems) or a "cell" (small complex nanosystems that function in larger quantities) that are revolutionary in nature and require a much longer term to develop. They utilize MEMS based manufacturing technologies.

Top Down Strategy: Generally speaking, the approach consisting in trying to build something small from something bigger. Richard Feynman suggested that large machines could be used to build smaller machines, which could themselves build smaller machines, and so on... In micro or nano-technology, the expression refers to all the methods and tools allowing us to build small components, starting from regular matter. It is opposed to the "bottom up strategy".

Top-down: Molding, carving and fabricating small materials and components by using larger objects such as our hands, tools and lasers.

Top-down Approach: The synthesis of a nanostructured material or a nanoscopic particle, nanocrystal by 'sculpting' it from the macroscopic piece of unstructured material (bulk). Common techniques used for this purpose are lithography, etching, mechanical milling, engineering using atomic force microscope (which can be used to 'scratch' the surface leaving nanoscopic channels and structures behind)...

Top-Down Nanotechnology: Engineers taking existing devices, such as transistors, and making them smaller. Top-down or mechanical nanotechnology will have the greatest impact on our everyday lives in the near future.

Top-Down: Molding, carving and fabricating small materials and components by using larger objects such as our hands, tools and lasers.

Topography: Refers to the contour or shape of a surface. The various semiconductor manufacturing processes like CVD, epitaxial growth, etching, etc. have an effect on the topography of a silicon wafer.

Topoisomerase (DNA topoisomerase): Enzyme that makes reversible cuts in a double helical DNA molecule for the purpose of removing knots or unwinding excessive twists.

Torque: A force that produces rotation. It is also the measure of the value of a force and is the product of the force and the perpendicular distance from the line of action of the force to the axis of rotation. The SI unit is Newton-meter.

Torr: A unit of pressure. It is equivalent to a pressure exerted by a column of 1 mm of Hg (mercury). One atmospheric pressure = 760 torr. $1 - 10^{-3}$ Torr is low vacuum and $10^{-3} - 10^{-9}$ torr is high vacuum.

Torsional Micromirrors: A micromirror that changes its position on the application of a potential. Increasing the driving voltage results in larger rotation angles. Its applications include digital projection displays, printers, and optical communications equipment.

Total Internal Reflectance Fluorescence Microscopy TIR-FM: Is based on the generation of an evanescent wave generated by total internal reflection at the boundary between media of differing refractive indices. The evanescent wave propagates in a direction normal to the interface for a short distance. Thus, it is useful for excitation of molecules in the vicinity of the surface- permitting membrane binding/ adsorption studies without the need to separate bulk phase ligand. Evanescent waves have been generated utilizing two- photon excitation with an accompanying decrease in the sensing depth.

Total Internal Reflection: The reflection that occurs when light, in a higher refractive-index medium, strikes an interface, with a lower-refractive index medium, at an angle of incidence greater than an angle specific to the two media.

Toughness: A measure of the amount of energy absorbed by a material as it fractures, indicated by the total area under the material's tensile stress-strain curve.

Transducer Electronic Data Sheet (TEDS): A set of electronic data needed by software for mapping between transducers and describe itself to the network. A TEDS is a persistent memory that specifies one or more transducers (sensors, actuators, or digital input or output devices) and their triggering, signal conditioning, and signal conversion. The TEDS is part of a STIM.

Transducer: A measuring device that provides an output quantity having a given relationship to the input quantity. E.g., (a) thermocouple; (b) current transformer; (c) electropneumatic converter.

Transduction (self-generating or modulating): The conversion of the signal to be measured into another, more easily accessible form. Source of energy for transmission of the sensor signal.

Transduction Mode (direct or indirect): How the sensor acquires the desired information from the material. In general, this parameter is an indication of the ability of the sensor signal to provide information regarding a material property or state of interest.

Transformation: A statistical technique to modify all the values of a variable by using a mathematical operation. For example, a transformation of a group of numbers occurs by multiplying by a constant or taking the natural logarithm or similar operations. The data can now be viewed in a different perspective to find new information that was previously not noticeable.

Transformer: Device using magnetically linked inductors to change AC voltage level.

Transhuman: Someone actively preparing for becoming posthuman. Someone who is informed enough to see radical future possibilities and plans ahead for them, and who takes every current option for self-enhancement.

Transhuman: Someone actively preparing for becoming posthuman. Someone who is informed enough to see radical future possibilities and plans ahead for them, and who takes every current option for self-enhancement.

Transhumanism: Philosophies of life (such as Extropianism) that seek the continuation and acceleration of the evolution of intelligent life beyond its currently human form and human limitations by means of science and technology, guided by life-promoting values.

Transient Enhanced Diffusion (TED): The diffusion of dopant ions due to the interaction of defects, like silicon interstitials and vacancies, with dopants during annealing after ion implantation. The dopant diffusivities are enhanced by many orders of magnitude but the increase is transient since the defects ultimately diffuse into the bulk or get destroyed at the surface.

Transient Response: The response of the sensor to a step change in the measurand.

Transient Voltage Suppression (TVS) Diodes: These are avalanche diodes designed specifically to protect other semiconductor devices from electrostatic discharges. Their p-n junctions have a much larger cross-sectional area than those of a normal diode, allowing them to conduct large currents to ground without sustaining damage

Transistor Outline (TO) Package: Standard mechanical packages used primarily for discrete transistors and low-pin-count ICs.

Transistor: (1) Semiconductor device used for amplification and switching. (2) The basic element in an integrated circuit. An on/off switch (consisting of three layers of a semiconductor material) that consists of a source (where electrons come from), a drain (where they go) and a gate that controls the flow of electrons through a channel that connects the source and the drain. There are two kinds of transistor, the bipolar transistor (also called the junction transistor), and the field effect transistor. (3) An active semiconductor device with three electrodes that may be used either as an amplifier or a switch.

Transition State Theory: Any of several theories that give approximate descriptions of chemical *reaction* rates based on the *PES* of the system, and in particular, on the properties of two *potential wells* and a *transition state* between them.

Transition State: At the saddle point of a *col* linking two *potential wells*, the direction of maximum negative curvature defines the reaction coordinate; the transition state is a hypothetical system of reduced dimensionality, free to move only on a hypersurface perpendicular to the reaction coordinate at its point of maximum energy.

Translation: Conversion of genetic information from messenger RNA (mRNA) into protein. Uniform motion of a body in a straight line.

Translational Symmetry: As it applies to superconductivity, translational symmetry is where the process of charge transfer is repeated exactly as the charge carriers (paired electrons) traverse the solid. In a normal conductor, latent heat continuously vibrates the atomic lattice, deflecting mobile free electrons and preventing "perfect" translational symmetry. In a superconductor this scattering tendency is overcome.

Transmission Electron Microscope (TEM): (1) A microscope that produces an image by using electron beams to transmit (pass through) the specimen, making examination of internal features at high magnifications possible. (2) A method of producing images of a sample by illuminating the sample with electronic radiation (in a vacuum), and detecting the electrons that are transmitted through the sample.

Transmission: Refers to system for carrying electric power at voltages above 100,000 volts.

Transponder: An instrument used on communications satellites that receives a signal from a station on Earth at one frequency, amplifies it, and shifts it to a new frequency.

Transtegumental: Crossing or passing through the skin or covering of a body.

Travelling Wave Tube: A hollow tube that is designed to fit a wave of a particular frequency. The width of the tube depends on the frequency of the signal. This results in a travelling wave.

Trench: An etched region filled with oxide or doped silicon that acts as an isolation region between adjacent devices. It is also used to fabricate capacitors.

Triac: A triac is an electronic component equivalent to two silicon controlled rectifiers joined end to end or back to back with their gates connected together. This results in a two-directional electronic switch which can conduct current in both directions when it is triggered (turned on). It can be triggered by either a positive or a negative voltage. The device turns off again when the current through it drops below a certain threshold value, such as the end of a half-cycle of alternating current. This makes the triac a convenient switch for AC circuits. Applying a trigger pulse at a controllable point in an AC cycle allows one to control the percentage of current that flows through the triac. Lower power triacs are used in lots of applications, especially domestic electrical appliances, such as light dimme

Tribology: (1) Study of friction, wear and lubrication of interacting surfaces. (2) The science and technology of two interacting surfaces in relative motion and of related subjects and practices. The popular equivalent is friction, wear, and lubrication in surfaces sliding against each other, as in bearing and gears.

Trichlorosilane (TCS): Stands for $SiHCl_3$, which is a liquid used in epitaxial silicon growth or polysilicon.

Trigonometry: An area of mathematics involving triangles.

Trigonometric calculations use the relationships between the sides and the angles of triangles to calculate position, distance, speed, and many other things.

Triple Bond: A double bond is formed when a pi bond is superimposed on a *single bond*; adding a second pi bond results in a triple bond. The two pi bonds have perpendicular nodal planes, and their sum has roughly cylindrical symmetry, permitting rotation in much the same manner as a single bond.

Triplet: An electronic state of a molecule in which two spins are aligned. This term is derived from spectroscopy: a system of two aligned spins has three possible orientations with respect to a magnetic field; each has a different energy, resulting in sets of three field-dependent spectral lines.

Troposphere: The lowest layer of the atmosphere. The troposphere has distinctive winds and cloud formations, and it has a very marked drop in temperature with altitude. It is 10 - 16 km from the surface of the Earth.

TSA Tyramide Signal Amplification: Was developed by NEN (now a part of PerkinElmer Corporation) and licensed to Molecular Probes for in- cell and in-tissue applications, permits significant amplification of the detectability of targets by a horseradish peroxidase (HRP)– mediated scheme. In the TSA method, the labeled tyramide becomes covalently linked to tyrosine residues in or near the target.

T-Spice: A version of SPICE developed by Tanner Research Labs and is part of MEMS Pro.

TT&C: An acronym for the satellite subsystem Telemetry, Tracking, and Control. TT&C refers to the brain of a satellite and its operating system. TT&C is the satellite's method for storing and analyzing the data it collects, and controlling its various systems. It also logs every activity of the satellite, receives information from the ground station, and takes care of any general upkeep, or "housekeeping", the satellite needs to do.

TTL (or T^2L): Transistor-Transistor Logic. A bipolar technology where a transistor output is connected directly (rather than through a resistor or diode) to a transistor input of the next stage.

Tubeologist: Someone who knows their nanotubes inside and out, such as David Tom·nek [uhf]

Tubeologist: Someone who knows their nanotubes inside and out, such as David Tom·nek

Tungsten Hexaflouride (WF_6): A toxic liquid used in CVD of tungsten interconnects. The contact holes in ICs are filled with tungsten using WF_6.

Tungsten-bronze: A nebulous term used to describe alkali metal tungstenates, vanadates, molybdates, titanates and niobates. The term was originally coined to describe Na_xWO_3 compounds; the crystals of which look much like the copper-tin alloy known as bronze. There have been reports of superconductivity as high as 91K for a surface-doped sodium tungsten-bronze. This material was the first high-temperature superconductor discovered that does not contain any copper.

Tunneling: A classical particle or system could not penetrate regions in which its energy would be negative, that is, barrier regions in which the *potential energy* is greater than the system energy. In the real world, however, a *wave function* of significant amplitude may extend into and beyond such a region. If the wave function extends into another region of positive energy, the barrier is crossed with some probability; this process is termed tunneling (since the barrier is penetrated rather than climbed).

Turing Test: Turing's proposed test for whether a machine is conscious (or intelligent, or aware): we communicate via text with it and with a hidden human. If we can't tell which of our partners in dialogue is the human, we say the computer is conscious. [AS]

Turnaround Time: Completion time to manufacture a product.

Turn-on-voltage: Applied voltage required to produce conduction in a diode.

Twin Well: A process where both a p-type region and an n-type region are made adjacent to enable the fabrication of pmos and nmos devices in the n-well and p-well, respectively.

Twinning: A defect in which the crystal grows in two different directions from a plane.

Type II Diabetes: A disease that affects the body's ability to use sugars, starches, fats and proteins and is not genetically transmitted as Type I diabetes. With Type II diabetes, too much insulin is produced in the body that prohibits glucose from entering the muscle cells. In turn, glucose rises to abnormal levels in the blood. Type-II diabetics are non-insulin dependent.

U

Ubiquitin: Small, highly conserved protein present in all eucaryotic cells that becomes covalently attached to lysines of other proteins. Attachment of a chain of ubiquitins tags a protein for intracellular proteolytic destruction in a proteasome.

Ubiquitous Computing: Also known as "embodied virtuality", "smart environment" and "ambient intelligence". Computers that are an integral, invisible part of people's lives. In some ways the opposite of virtual reality, in which the user is absorbed into the computational world. With ubiquitous computing, computers take into account the human world rather than requiring humans to enter into the computer's methods of working.

ULPA Filter (Ultra-Low-Particle Air filter): An air filter that removes 99.999% of particles that are 0.1 µm or larger. Only 1 particle 0.1 µm in size out of 100,000 ones passes through the ULPA filter.

ULSI (Ultra Large Scale Integration): The placing more than one million circuit elements on a single chip. The same as VLSI and used more often in Japan.

ULSI: Ultra large scale integration; a chip with over 1,000,000 components.

Ultraconductor: Materials known as ultraconductors™ display room-temperature resistance many orders of magnitude lower than the best metallic conductors. Examples of these materials include oxidized atactic polypropylene (OAPP) and other polymers. Since ultraconductor™ is a colloquial term, these materials might better be described as "hyperconductors". The Meissner effect cannot be confirmed in them, but strong (giant) diamagnetism *is* in evidence. Some of them may actually find acceptance in high-current applications ahead of superconductors as a result of their low losses at ambient temperatures and pressures.

Ultrasonic Bonding: A bonding technique that utilizes ultrasonic energy and pressure to form bonds.

Ulysses: A NASA space probe that is studying the sun. In its passes over the polar regions of the sun in 1994 and 1995, Ulysses revealed the existence of fast solar winds coming from the poles. Ulysses will return to the sun in the year 2000 after making a slingshot around Jupiter.

***Un situ* Hybridization:** Technique in which a single-stranded RNA or DNA probe is used to locate a gene or an mRNA molecule in a cell or tissue.

Uncertainty Principle: A corner stone of quantum theory. It states that the position of the particle and its momentum (velocity) cannot be simultaneously measured with arbitrary precision—the more precisely one measures the position of the particle, the more he perturbs its velocity. Roughly, the product of the uncertainties in position and momentum must be larger than the Planck constant. The principle is also known as Heisenberg's uncertainty principle after German physicist and

Nobel Prize laureate Werner Heisenberg. A consequence of this principle is that the matter at the scale of nanometer and smaller (atoms) cannot be manipulated with arbitrary precision since this is prohibited at the fundamental level by the Quantum Theory. Heisenberg stated that atoms *'form a world of potentialities or possibilities rather than one of things or facts'*.

Undressing: The process by which a quasiparticle becomes more like a bare (normal) particle. It is theorized this may be a driving force behind superconductivity, as undressed electrons are significantly lighter and can, thus, conduct current more readily.

Unimolecular: Occurring to or within a single molecule; like *intramolecular*, but can refer to fragmentation reactions.

Unipolar: FET devices where current passes only through one type of semiconductor material (P or N) as it flows from input to output.

Unit Cell: A unit cell is the smallest assemblage of atoms, ions, or molecules in a solid, beyond which the structure repeats to form the 3-dimensional crystal lattice. The basic structural unit of a crystal structure, defined in terms of atom (or ion) positions within a parallelepiped volume.

Universal Assembler: This is a term that was introduced by Eric Drexler. In his terminology, an assembler is enzyme or catalyst that brings two molecules together and connects them. DNA can be used to program the assembly of a wide variety of proteins. Drexler imagines a universal assembler to be something that could be programmed to fabricate anything. The following quote was taken from Eric

Drexler's book Engines of creation in the section where he discusses universal assemblers.

These second-generation nanomachines - built of more than just proteins - will do all that proteins can do, and more. In particular, some will serve as improved devices for assembling molecular structures. Able to tolerate acid or vacuum, freezing or baking, depending on design, enzyme-like second-generation machines will be able to use as "tools" almost any of the reactive molecules used by chemists - but they will wield them with the precision of programmed machines. They will be able to bond atoms together in virtually any stable pattern, adding a few at a time to the surface of a workpiece until a complex structure is complete. Think of such nanomachines as assemblers.

Because assemblers will let us place atoms in almost any reasonable arrangement (as discussed in the Notes), they will let us build almost anything that the laws of nature allow to exist. In particular, they will let us build almost anything we can design - including more assemblers. The consequences of this will be profound, because our crude tools have let us explore only a small part of the range of possibilities that natural law permits. Assemblers will open a world of new technologies.

Advances in the technologies of medicine, space, computation, and production - and warfare - all depend on our ability to arrange atoms. With assemblers, we will be able to remake our world or destroy it. So at this point it seems wise to step back and look at the prospect as clearly as we can, so we can be sure that assemblers and nanotechnology are not a mere futurological mirage.

Uses raw atoms and molecules to construct consumer goods, and is pollution free. Can be programmed to build anything that is composed of atoms and consistent with the rules of chemical stability. Eric Drexler talks about these assemblers as nanorobots with telescoping manipulator arms that are capable of picking up individual atoms, and combining them however they are programmed.

Universal Constructor: A machine capable of constructing anything that can be constructed. The physical analog of a "universal computer", which can perform any computation.

Unsaturated: Describes a molecule that contains one or more double or triple carbon-carbon bonds, such as isoprene or benzene.

Uplift: To increase the intelligence and help develop a culture of a previously non- or near-intelligent species. [From the Uplift novels by David Brin] [AS]

Uplinked: The process by which a ground station sends information and instructions up to a satellite. Often, scientists on Earth will have to send instructions to a satellite's on board computer. The ground station translates its computer information into radio waves and sends those waves up to the satellite via an antenna, usually in the form of a dish. Another antenna, on the satellite, picks up the radio waves sent from the Earth, and translates them back into a form that the satellite's on board computers can understand.

Upload: (1) To transfer the consciousness and mental structure of a person from a biological matrix to an electronic or informational matrix (this assumes that the strong AI postulate holds). The term "Downloading" is also sometimes used, mainly to denote transferring the mind to a slower or less spacious matrix. (2) The resulting infomorph person. [The origin of the term is uncertain, but obviously based on the computer technology term 'uploading' (loading data into a mainframe computer).

Usenet: Interlinked bulletin boards available via Internet and commercial on-line services.

Utility Fog: [AKA: Polymorphic Smart Materials] Objects formed of "intelligent" polymorphic (able to change shape) substances, typically having an octet truss structure. Concept concieved by Dr. J. Storrs Hall. Imagine a microscopic robot. It has a body about the size of a human cell and 12 arms sticking out in all directions. A bucketfull of such robots might form a 'robot crystal' by linking their arms up into a lattice structure. Now take a room, with people, furniture, and other objects in it — it's still mostly empty air. Fill the air completely full of robots. The robots are called Foglets and the substance they form is Utility Fog, which may have many useful medical applications. And when a number of utility foglets hold hands with their neighbors, they form a reconfigurable array of 'smart matter.'" Copyright Dr. J. Storrs Hall Research Fellow of the Institute for Molecular Manufacturing.

UV Embossing: Cold embossing uses UV light to polymerize and solidify materials after forming by the stamp. The starting materials are thin films of UV-curable materials. Typical materials may be monomers, or inorganic-organic hybrid polymers, which are polymerized by UV to polymers with desirable optical properties. The films may be deposited onto silicon, gallium arsenide, or similar wafers. The UV-transmissive patterned master template is pressed against the film, imprinting the pattern, which is UV cured before pressure is released. UV embossing allows accurately aligning and overlaying multiple levels of features. UV embossing has a clear advantage over hot embossing in allowing multi-level applications, since the first level is polymerized after exposure, making it irrevocably formed and dimensionally stable, regardless of subsequent layer applications and exposures. UV embossing is a low pressure, room temperature process, with little shrinkage of the pattern. It also allows partial UV casting, by limiting the illumination to maintain clear areas on the wafer. Cold embossing may be done on one or both sides of the wafer. Double-sided embossing allows micromolding.

UV (Ultraviolet Rays): Radiation that has a wavelength between 4 nm and 480 nm. Most lithographic equipment for IC and MEMS fabrication use UV radiation.

V

Vacancy Diffusion: The migration of an atom from a typical lattice position to a vacancy lattice site. The migrating atom has to be of the same size or has to be a substitutional atom. See also Interstitial Diffusion.

Vacuum: An environment with pressure significantly lower from the atmospheric pressure. Generally, the pressure is due to the force exhibited by atoms and molecules colliding with the object immersed in the environment. So, the *perfect* vacuum is the space void of matter. Even the interstellar space does not represent the perfect vacuum. Ultra-high vacuum environments are produced in laboratories using the special equipment (steel chambers which are evacuated by highly efficient pump(s)).

Valence Band: The electron energy band that contains the valence electrons in solid materials.

Valence Electrons: (1) Electrons that can participate in bonds and in chemical reactions; lone-pair electrons are valence electrons, although not participating in a bond. The electrons in the outermost occupied electron shell, that participate in interatomic bonding. (2) The electrons in the outermost occupied electron shell, that participate in interatomic bonding.

Valence: In covalent compounds, the valence of an atom is the number of bonds it forms to other atoms.

Van der Pauw Method: Away of measuring the sheet resistance of a wafer using a four-probe system.

Van der Waals Bond: A secondary, permanent or induced, interatomic bond between adjacent molecular dipoles. Any of several intermolecular attractive forces not resulting from ionic charges; in this volume, only London dispersion forces are described by this name. Descriptions of the potential energy of nonbonded interactions follow the convention of expressing van der Waals attractive forces and overlap repulsion forces as a single *van der Waals potential*.

Van der Waals Force: Any of several intermolecular attractive forces not resulting from ionic charges; in this volume, only London dispersion forces are described by this name. Descriptions of the potential energy of nonbonded interactions follow the convention of expressing van der Waals attractive forces and overlap repulsion forces as a single *van der Waals potential*.

Vapor Plating: A vacuum process, usually less than 10^{-10} Torr (mm of Hg), where metal(s) are vaporized through thermal agitation, then recrystallized on cooler surfaces, generally the material to be coated. It's also referred to as evaporation.

Vapor Pressure: The pressure in the space above a liquid in a closed container. Compounds with high vapor pressures are volatile. The boiling point of a liquid is the temperature at which the vapor pressure of the liquid equals the external pressure.

Vapor Priming: A process of exposing wafers to a vapor of HMDS to prime the wafer surface prior to photoresist coating. It improves photoresist adhesion, consuming less HMDS since a monolayer of HMDS is formed.

Variable Cost: A cost that increases or decreases in proportion to the level of output. The variable cost per unit may remain level over different production volumes.

Variance: The square of the standard deviation. See Standard Deviation.

Vascular Graft: An implant of a new vessel in the vascular system.

Vascular System: The heart, blood, and network of blood vessels that lead to and from the heart.

Vasculocyte: In medical nanorobotics, a theorized (nanorobotic) device capable of performing repairs of an injured vascular luminal surface.

Vasculography: A physical description (and mapping) of the human vascular system.

Vasculoid Removal: The procedures outlined above for vasculoid installation are designed to be fully reversible at every step. Thus, vasculoid removal may be accomplished by reversing the installation procedure, except that patient preparation should be performed first. If the patient has worn his or her vascular appliance for a considerable time, heart muscles, venous valves, splenic filtration systems and the erythroid marrow may have significantly atrophied, requiring remedial cellular repair. Homologous formed blood elements such as RBCs must also be premanufactured for postoperative infusion prior to vasculoid removal, and any genetically modified endothelial cells must be reprogramed back to their original state.

Vasculoid: Vasculoid is a single, complex, multisegmented nanotechnological medical robotic system capable of duplicating all essential thermal and biochemical transport functions of the blood, including circulation of respiratory gases, glucose, hormones, cytokines, waste products, and cellular components. See Vasculoid: A Personal Nanomedical Appliance to Replace Human Blood Robert A. Freitas Jr. & Christopher J. Phoenix. April 2002

VC (Venture Capitalist): A person or organisation that invests in promising companies during early development stages.

Vector: In cell biology, an agent (virus or plasmid) used to transmit genetic material to a cell or organism.

Vegetal Pole: The end at which most of the yolk is located in an animal egg. The end opposite the animal pole.

Velocity: The measure of how fast an object is moving in a particular direction.

Ventral: Situated toward the belly surface of an animal.

Ventricular Tachycardia: A condition that occurs when ventricles (a pair chambers in the heart) beat at such an abnormally rapid rate that they are unable to fill completely with blood, impairing the ventricles' ability to pump oxygenated blood throughout the body.

Verification Process: The determination, by review of quality evidence, that the results from a given project activity satisfy conditions imposed at the start of that activity. The act of confirming, substantiating, and assuring that an activity or condition has been implemented in conformance with specified requirements.

Verification: Tests and analyses to be performed during the design, development, assembly, integration, and operational phases of a measurement system to assure that specified requirements have been met. It includes all subsystem and system tests done at the functional level.

Vernier Resolution: The measuring capacity of the smallest measuring instrument in a measuring device.

Vertical Foundries: Foundries that include all elements needed to produce finished products. Example companies include Standard MEMS and TMP.

Vertical Integration: The degree to which a single firm controls the production and distribution process from initial resources to the end customer. It also refers to the combination of two or more stages in a production chain.

Vesicle: Small, membrane-bounded, spherical organelle in the cytoplasm of a eucaryotic cell.

VHDL (Very high-speed integrated-circuit Hardware Description Language): It is a standardized example of a hardware description language that has the capability of specifying function and behavior of circuits.

Via: A path filled with conducting material between two metallic layers in an IC.

Viking: A NASA space probe that gave us the first close look at Mars. Viking was made up of an Orbiter and a Lander, which was sent down to Mars to study the Martian soil and atmosphere.

Virtual Microscopy: The technique of digitizing a glass microscope slide with any commonly used objective magnification producing a digital virtual microscope slide with diagnostic image quality. A virtual microscope slide should have quality and resolution that is similar to looking at the original glass slide on a microscope.

Virtual Nanomedicine: Using VR to perform surgery and other functions inside the body.

Virtual Reality System: A combination of computer and interface devices (goggles, gloves, etc.) that presents a user with the illusion of being in a three dimensional world of computer-generated objects.

Virus: A structure consisting of pieces of genetic material (nucleic acid) surrounded by a protective coating made of proteins or proteins combined with carbohydrate or lipid components. Protein coatings of smallest viruses are very regular and can be precisely described mathematically. Small viruses are most often shaped like icosahedrons, but other regular shapes appear also. For example, mosaic tobacco virus looks like a cylinder.

A parasite (consisting primarily of genetic material) that invades cells and takes over their molecular machinery in order to copy itself. A small replicator consisting of little but a package of DNA or RNA which, when injected into a host cell, can direct the cell's molecular machinery to make more viruses. Particle consisting of nucleic acid (RNA or DNA) enclosed in a protein coat and capable of replicating within a host cell and spreading from cell to cell. Often the cause of disease.

A small replicator consisting of little but a package of DNA or RNA which, when injected into a host cell, can direct the cell's molecular machinery to make more viruses.

Viscoelasticity: A type of deformation exhibiting the mechanical characteristics of viscous flow and elastic deformation.

Viscosity (symbol is ç): The ratio of the magnitude of an applied shear stress to the velocity gradient that it produces; in other words: a measure of a noncrystalline material's resistance to permanent deformation.

Viscosity: The property of resistance to flow in a fluid (a gas or liquid).

Viscosity: The ratio of the magnitude of an applied shear stress to the velocity gradient that it produces; in other words: a measure of a noncrystalline material's resistance to permanent deformation.

Visible Spectrum: Visible light makes up only a small part of the *electromagnetic spectrum*. The visible light spectrum can be divided into different wavelenghts of light. The wavelength of the light determines the color of that light. The light spectrum goes from violet to red where red is the longest wavelength.

Vitamins (Engineering): In machine replication theory, vitamin parts are components of a self-replicating machine which the machine is incapable of producing itself, therefore these vital parts must be supplied from an external source.

Vitrification: During firing of a ceramic body, the formation of a liquid phase that becomes a glass-bonding matrix upon cooling.

VLSI (Very Large Scale Integration): The fabrication technology for concentrating many thousands of semiconductor devices into a single IC.

Strictlyspeaking, VLSI is in the range of >10,000 transistors but less than 1,000,000.

Vmax: Maximum rate of an enzymatic reaction when a substrate is present at saturation levels.

VOC (Volatile Organic Compound): Any compound of carbon (excluding carbon-monoxide), carbon-dioxide, carbonic acid, metallic carbides or carbonates, and ammonium-carbonate, that participates in atmospheric photochemical reactions. These compounds have the potential as refrigerants, aerosol propellants, fire extinguishers, blowing agents, and solvents. (Source: United Sates Environment and Protection Agency).

Volatile: The relative inclination of a substance to become a gas from a liquid or solid form.

Volitional Normative Model of Disease: In medical nanorobotics, disease is said to be present in a human being upon either (1) the failure of optimal physical (e.g. biological) functioning, or (2) the failure of desired (by the patient) functioning.

Voltage [V]: Potential difference between two points: energy to move a 1-C charge through a 1-V potential difference is 1-J.

Voltage Regulator Device: A chip that controls voltage delivered to a load.

Vombinatorial Biology: Involves genetic manipulation of bacteria and fungi that produce complex natural products. This technology includes construction of large libraries of recombinant microbes capable of generating novel organic molecules and engineering secondary metabolite biosynthetic pathways to modify valuable biologically active microbial metabolites. *Combinatorial materials design:* Uses computing power (sometimes together with massive parallel experimentation) to screen many different materials possibilities to optimize properties for specific applications (e.g., catalysts, drugs, optical materials).

Von Neumann Machine: (pronounced von noi-man) A machine which is able to build a working copy of itself using materials in its environment. This is often proposed as a cheap way to mine or colonize the entire solar system or galaxy. An early fictional treatment was the short story "Autofac" by Philip K. Dick, published in 1955, which actually seems to precede John von Neumann's original paper about self-reproducing machines (von Neumann, J., 1966, The Theory of Self-reproducing Automata, A. Burks, ed., Univ. of Illinois Press, Urbana, IL.).

VON Neumann Probe: A von Neumann Machine able to move over interstellar or interplanetary distances and to utilize local materials to build new copies of itself. Such probes could be used to set up new colonies, perform megascale engineering or explore the universe. [AS]

Vone Substitutes: Synthetic or natural materials for the replacement of bones or bone tissue. They include hard tissue replacement polymers, natural coral, hydroxyapatite, beta-tricalcium phosphate, and various other biomaterials. The bone substitutes as inert materials can be incorporated into surrounding tissue or gradually replaced by original tissue.

Vortices (plural of vortex): Swirling tubes of electrical current induced by an external magnetic field into the surface of a superconducting material that represent a topological singularity in the wavefunction. These are particularly evident in Type 2 superconductors during "mixed-state" behavior when the surface is just partially superconducting. Superconductivity is completely suppressed within these volcano-shaped structures. Recent research suggests that flux vortices may NOT possess quantum values (equal to multiples of Planck's constant divided by 2 times electron charge). But may instead have but a tiny fraction of the basic unit of magnetism. The movement of vortices can produce a pseudo-resistance and, as such, is undesirable. While superconductivity is a "macroscopic" phenomenon, vortices are a "mesoscopic" phenomenon.

Voyager: Launched in 1977, NASA's twin Voyager *space probes* studied the outer planets before continuing on into deep space. The Voyager probes carried into space a record called "Sounds of Earth," bearing messages and pictures from our planet.

Wafer Boat: A quartz or Teflon container used to transport wafers from fabrication facilities to a clean room. Most boats have 25 or 50 slots.

Wafer Fabrication: The process of creating ICs on the surface of wafers.

Wafer Sort: The primary wafer-level electrical test to verify that each die functions properly at the completion of wafer fabrication.

Wafer Yield: The ratio of the number of wafers that have successfully completed the wafer fabrication process by the number of wafers initially put through the wafer fabrication process.

Wafer: A round disc of semiconductor material used to manufacture semiconductor devices in wafer fabrication, and available in a variety of diameters, e.g. 75 mm (3 in.), 100 mm, 125 mm, 150 mm (6 in.), 200 mm, and 300 mm (12 in.). A wafer contains several hundred die.

Waste Recycling: An advanced nanotechnology might be able to build filters that dynamically modify themselves to attract the contaminant molecules detected by the air and water quality sensors. Once attached to the filter, the filter could in principle move the offending molecules to a molecular laboratory for modifications to useful or at least inert products. A swarm might implement such an active filter if it was able to dynamically manufacture proteins that could bind contaminant molecules. The protein and bound contaminant might then be manipulated by the swarm for transportation. With a sufficiently advanced nanotechnology it might even be possible to directly generate food by non-biological means. Then agriculture waste in a self-sufficient space colony could be converted directly to useful nutrition. Making this food attractive will be a major challenge

Water Deionization: A process of removing the ions using ion-exchange resins to obtain deionized water that is used in semiconductor fabs.

Watt (W): Unit of power. One watt is the power that, in 1 second, gives rise to an energy of 1 joule.

Wave Filter: An instrument that clarifies a signal by filtering out all other noise. A wave filter is useful for receiving communications in deep space where there are other stellar phenomena that emit frequencies too. These extra frequencies become part of the white noise always present in space. A wave filter will keep a satellite from being confused by this extra "noise" from space.

Wave Function: In quantum mechanics, a complex function extending over the configuration space of a system; its complex conjugate yields the probability density function, and other mathematical operations yield other physical quantities.

Wave Guide: An instrument that allows satellites to manipulate radio waves and redirect them within the satellite for processing. Wave guides are essential for any satellite performing a function concerning communications.

Waveguide: A material medium that confines and guides a propagating electromagnetic wave. In optical telecommunication, a waveguide is an elongated transmission line that consists of a solid dielectric filament, usually circular in cross section. E.g., an optical fiber.

Wavelength Routing: A concept developed by Cisco Systems where at each intermediate node between the end nodes, light coming in on one port at a given wavelength gets routed out of one and only one port. It allows carriers to leverage their core network technologies to support the massive growth of data services. It avoids wasting transmitted power, by channeling the energy transmitted by each node along a restricted path instead of letting it spread over the entire network.

Wavelength. The distance between two adjacent wave crests. Visible light and X-rays are both electromagnetic waves and differ from each other only in the length of the wave. The wavelength of visible light ranges from 400 to 700 nanometres while the wavelength of X-rays ranges from about 0.01 to 10 nanometres. The relatively long wavelength of visible light sets the limit of how small an image it can produce. For more information see Electromagnetic radiation (Back to basics, Australian Academy of Science).

WDM Source Laser: A laser that tests power and spectral sensitivity of active components, passive components, and other WDM building blocks.

WDM: In optical-fiber communications, a technique by which two or more optical signals having different wavelengths may be simultaneously transmitted in the same direction over one fiber, and then be separated according to wavelength at the output end.

Weather Satellite: A type of satellite used to give meteorologists information about the weather. Weather satellites can do things like take pictures of cloud cover, monitor threatening weather systems like hurricanes, measure temperatures of the air and the sea, and generally give meteorologists the information that they need to make weather predictions.

Weber: Unit of magnetic flux. One weber is a magnetic flux that, linking a circuit of 1 turn, would produce in it an electromotive force of 1 volt if it were reduced to zero at a uniform rate in 1 second.

Weight Percent (wt%): A way of expressing the concentration of a substance on the basis of weight (or mass) of the dissolved substance in the solution. Weight percent = weight of solute/weight of solution x 100. Concentration specification on the basis of weight (or mass) of a particular element relative to the total alloy weight (or mass).

Welding: The joining of two or more pieces of metal by fusing them together.

Western Blotting: Technique by which proteins are separated and immobilized on a paper sheet and then analyzed, usually by means of a labeled antibody.

Wet Etching: A process of removing material during semiconductor fabrication using wet-chemical agents.

Dry etching is preferred to wet etching in the semiconductor industry.

Wet Nanotechnology: The study of biological systems that exist primarily in a water environment. The functional nanometer-scale structures of interest here are genetic material, membranes, enzymes and other cellular components. The success of this nanotechnology is amply demonstrated by the existence of living organisms whose form, function, and evolution are governed by the interactions of nanometer-scale structures.

Wet Oxidation: The introduction of water vapor during oxidation to increase the growth rate. The oxide grown is not electrically stable.

Wheatstone Bridge: A bridge for measuring electrical resistances that consists of a conductor joining two branches in parallel, each branch having two resistors connected in series. The resistance of one of the four resistors can be found by this arrangement.

Whisker: A very thin, single crystal of high perfection which has an extremely large length-to-diameter ratio. Whiskers are used as the reinforcing phase in some composites.

White Biotechnology: An emerging field within modern biotechnology that serves industry. It uses living cells like moulds, yeasts or bacteria, as well as enzymes to produce goods and services. Living cells can be used as they are or improved to work as "cell factories" to produce enzymes for industry.

White Blood Cell (leucocyte): Nucleated blood cell lacking hemoglobin; includes lymphocytes, neutrophils, eosinophils, basophils, and monocytes.

Wide Bandgap Semiconductors: Wide Bandgap Semiconductors have applications in optoelectronic and power devices. Gallium Nitride (GaN) is of particular interest, as it is used to make blue LEDs and lasers. In the future, high brightness, long life white LEDs may replace incandescent bulbs in many situations. The next generation of DVD players, the Blue ray disk, uses GaN based lasers. Applications for high power and temperature devices are also interesting. Both Gallium Nitride and Silicon Carbide are robust materials well suited for such applications

Wild-type: Normal, nonmutant form of an organism; the form found in nature (in the wild).

Wind Imaging Interferometer (WINDII): In 1991, the Canadian WINDII was launched on NASA's Upper Atmosphere Research Satellite and has been providing global information on temperatures, wind patterns, and concentrations of important gases in the upper atmosphere.

Wire Bonding: A method to make electrical connections between aluminum bonding pads on the surface of the chip to the package lead terminals.

Wiring: The electrical circuitry on a wafer. Also called interconnect.

Within Wafer Non-Uniformity (WIWNU): A measure of the amount of film thickness variation across the entire wafer. It is used to evaluate equipment performance and compare different processes. WIWNU increases as wafer size increases.

Within-Die Non-Uniformity (WIDNU): The average amount of film thickness variation across a die. Overall film thickness variation of a film coated on top of a wafer is the sum of within-die non-uniformity (WIDNU) and within-wafer non-uniformity (WIWNU).

Wobble Motors: A low-cost alternative to electric motors. These motors use rotors with diameters between 120 and 500 μm, are robust, require no lubrication, have higher torque-to-weight ratios than conventional motors, and are inexpensive. Thus they can be used in miniature medical devices. Manufactured by Sarcos Inc., Utah.

Work In Process (WIP): The input materials still in a manufacturing process. In the semiconductor industry, it is the wafers still being subjected to various manufacturing processes until final packaging. It is measured in units of wafer or in dollar terms.

Work: Energy transferred by applying a force over a distance; lifting a mass does work against gravity, and stores gravitational potential energy.

Working Plates: Masks printed from master plates that are used for the production exposure of wafers. As these plates are subject to wear, they must be replaced periodically.

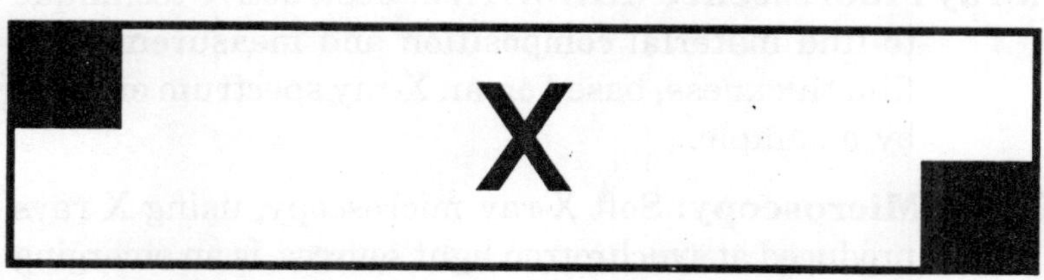

Xenografts: The live cells, tissues and organs used in xenotransplant procedures.

Xenopus Laevis (South African clawed toad): Species of frog (not a toad) frequently used in studies of early vertebrate development.

Xenotransplantation: Xenotransplantation involves the use of live animal cells, tissues and organs in the treatment or mitigation of human disease. Xenotransplantation has the potential to provide an alternative source of tissue for human transplantation. However, there are concerns that in crossing the species barriers, infectious agents which normally do not cause any problem in the original species may be changed or may adversely respond to the new species into which the infectious agent is transplanted.

X-ray Crystallography: X-ray crystallography involves firing X-rays through the crystal of a molecule to produce a diffraction pattern. This pattern provides information on the structure of that crystal. For example, X-ray crystallography helped scientists discover that the DNA molecule exists as a double helix. For more information see Crystallography 101 (Lawrence Livermore National Laboratory, USA) and X-ray crystallography (The British Biophysical Society, UK).

X-ray Fluorescence (XRF): A non-destructive technique to find material composition and measurement of film thickness, based on an X-ray spectrum emitted by a sample.

X-ray Microscopy: Soft X-ray microscopy, using X rays produced at synchrotron light sources, is an emerging biological imaging technique for the examination of intact, hydrated cells. The shorter wavelengths of X rays permit resolution 5 to 8 times better than that achieved by light microscopy, and the information obtained from the image contrast is highly quantitative in nature. Protein location can be determined at better than 50nm resolution in whole cells.

X-ray: Electromagnetic radiation with a wavelength greater than 10×10^{-9} m and less than 10×10^{-12} m. It has a smaller wavelength than an ultra-violet ray but more than a gamma ray.

Y

YBCO: An acronym for a well-known ceramic superconductor composed of Yttrium, Barium, Copper and Oxygen. This was the first truly "high temperature" ceramic superconductor discovered; having a transition temperature well above the boiling point of liquid nitrogen—a commonly available coolant. Its actual molecular formula is $YBa_2Cu_3O_7$, making it a "1-2-3" superconductor. YBCO compounds exhibit d-wave electron pairing. The patent for YBCO is held by Lucent Technologies. With the discovery of the ceramic superconductors, came a need for a classification system to describe structure types. The cuprate superconductors all have blocks of conducting CuO_2 planes, alternating with insulating, spacing and separating layers. This makes possible a systematic Naming Scheme that allows for identification and comparison. The scheme chosen uses four numbers. The first denotes the number of insulating layers between adjacent conducting blocks. The second represents the number of spacing layers between identical CuO_2 blocks. The third gives the number of layers that separate adjacent CuO_2 planes within the conducting block. And, the fourth is the number of CuO_2planes within a conducting block. Using the $TlBa_2Ca_2Cu_3O_9$ molecule depicted at left as an example, there is 1 insulating TlO layer, 2 spacing BaO layers, 2 separating Ca layers, and 3 conducting CuO_2 planes - making it a "1223" type. There are also occasionally letters of the alphabet in the 5th and 6th positions. Suffixes like "T*", "C", and "F" describe to researchers structure detail in the oxygen positions. To further complicate matters, several of the more popular structures have names that do NOT follow the 4-number scheme. For example, the compound $Y_1Ba_2Cu_3O_7$ is often referred to by the 3-number name "123". This delineates the number of metal atoms (the stoichiometry)— without regard to atom location—within the structure. Using the 4-number scheme, it would be classified a "1212".

Yeast: Common term for several families of unicellular fungi. Includes species used for brewing beer and making bread, as well as pathogenic species (that is, species that cause disease).

Yellow Room: The room in which wafers are exposed to ultraviolet light in an aligner. Fluorescent lights in the room have yellow filter tubes around them to block unwanted ultraviolet light from that source.

Yield Model: A mathematical probability model used to convert defect density to a projected yield.

Yield Strength: The stress required to produce a very slight yet specified amount of plastic strain; a strain offset of 0.002 is commonly used.

Yield: The percentage of the total number of die available on a wafer that function to specifications.

Yielding: The onset of plastic deformation.

Young's Modulus: A modulus relating tensile (or compressive) stress to strain in a rod that is free to contract or expand transversely in accord with its Poisson's ratio. The relevant measure of strain is the elongation divided by the initial length.

Z

Z Disc (Z line): Platelike region of a muscle sarcomere to which the plus ends of actin filaments are attached. Seen as a dark transverse line in micrographs.

Zener Diode: A P-N junction two-terminal, semiconductor device reverse-biased into the breakdown region. It offers high impedance when a voltage less than the breakdown voltage is applied. Current increases dramatically at voltages over the breakdown voltage and hence it is used as a voltage regulator.

Zener Diodes: Diodes that can be made to conduct backwards. This effect, called Zener Breakdown, occurs at a precisely defined voltage, allowing the diode to be used as a precision voltage reference. Some devices labelled as high-voltage Zener diodes are actually avalanche diodes. Two (equivalent) Zeners in series and in reverse order, in the same package, constitute a transient absorber (or *Transorb*, a registered trademark). They are named for Dr. Clarence Melvin Zener of Southern Illinois University, inventor of the device. Semiconductor diode that has a well-defined turn-on voltage for conduction in the reverse direction.

Zephyr: A common term used to describe lightweight fabrics. The Zephyr philosophy is *building compilers* *from parts*. Parts might include front ends, back ends, optimizers, and the glue that holds all these pieces together. You might even *generate parts automatically* from compact specifications.

If you *describe your intermediate forms* using Zephyr's Abstract Syntax Description Language (ASDL), we can generate data-structure definitions in *C, C++, Java, Standard ML, and Haskell*. Your IR can be serialized on disk and *freely exchanged* among compiler passes written in these languages; asdlGen creates the glue that holds your compiler together. You can also *visualize your IR* using our graphical browser.

To *compile code for multiple targets*, the state of the art is to use a code-generator generator, which lets you specify a translation from an intermediate code to a target instruction set. We're trying to separate the intermediate code from the instruction set, so we can develop *machine descriptions that can be reused*, not just with compilers, but also with other tools. The idea is to *generate the machine-dependent parts* from descriptions of instructions' semantics, of binary representations, or of other properties. Zephyr's Computer Systems Description Languages (CSDL) let you *describe as much or as little as you need* for your application.

Getting high-quality code requires optimization. Zephyr's very portable optimizer (vpo) provides *instruction selection*, *instruction scheduling*, and classical *global optimization*. Vpo operates on programs represented as register-transfer lists (RTLs) in SSA form. All vpo optimizations are "plug and play," and it is easy to add new ones. Vpo provides a level playing field for *evaluating different optimizations*.

Zeptosecond: One-billion-trillionth of a second, or 10^{-21} second. Because nuclear movement takes place so quickly, scientists would need a pulse of light lasting just one zeptosecond to observe them. Johns Hopkins University

Zero Offset: The output of a sensor at zero input for a specified supply voltage or current. The zero offset of the powermeter relates to the torque part of this equation. When there is no force being applied to the powermeter, there is still a small amount of torque being generated. This is because of the properties of the metal from which the powermeter is made; as temperature changes, the metal either expands or shrinks. The bolts attached to the powermeter and the plastic lid also slightly deform the metal. The strain gauges (shown at right) in the powermeter pick up these metal deformations, and they register on the Powercontrol as a frequency — called the zero offset frequency. If the zero offset is not set before training, this 'background' frequency is not correctly subtracted from the frequency generated by applied torque, and the calculated power will be incorrect.

If the deviation of the slope and the zero offset is very small, the measurement of power will be very precise (up to +/-1%).

The zero offset should be reset in the following cases:

- After remounting the powermeter, or after changing the chainrings or crank.After cycling for a short time (20 km) one should carry out a second calibration of the zero offset.
- If there are large fluctuations of temperature (about ±20 degrees).

- At the beginning of every ride that you wish to record.
- Before very important measurements.

Setting the zero offset frequency:

- Switch the Powercontrol on by pressing the Mode key
- Switch the Powermeter on by rotating the cranks backwards at least twice. Then apply forward torque by pressing down hard on the right pedal. Release the pedal, and rotate the cranks backwards a few more times.
- Let go and *make sure that there is no tension on the chain*, or the zero offset frequency will be too high, and power readings will be too low.
- Go into calibration mode by pressing the Set and Mode key at the same time.
- Wait for a few seconds until the number on the right side of the display (under the 'Set' heading) is nearly constant (+/- 5 Hz). The number to the right should not deviate more than +/-10 Hz from the number on the left (under the 'MANU' heading) if no force is exerted on the pedals.

Press the Set key to accept the number on right as the correct zero frequency (zero offset). This number will move to the left (under "Manu" for Manual) and will be stored as the zero frequency for calculating the power. The number in the middle is not important. This is the zero offset that the computer proposes. You can accept this by pressing the Pro key.

Zero Sum: A term used to describe a situation in which one entity can gain only if other entities suffer an equal loss; for example, a private poker game. In his earliest work, von Neumann made a striking discovery. He found that if poker players maximize their rewards, they do so by bluffing; and, more generally, that in many games it pays to be unpredictable. This was not qualitatively new, of course — baseball pitchers were throwing change-up pitches before von Neumann wrote about mixed strategies. But von Neumann's discovery was a bit more than just that. He discovered a unique and unequivocal answer to the question "how can I maximize my rewards in this sort of game?" without any markets, prices, property rights, or other institutions in the picture. It was a very major extension of the concept of absolute rationality in neoclassical economics. But von Neumann had bought his discovery at a price. The price was a strong simplifying assumption: von Neumann's discovery applied only to zero-sum games.

For example, consider the children's game of "Matching Pennies." In this game, the two players agree that one will be "even" and the other will be "odd." Each one then shows a penny. The pennies are shown simultaneously, and each player may show either a head or a tail. If both show the same side, then "even" wins the penny from "odd;" or if they show different sides, "odd" wins the penny from "even". Here is the payoff table for the game.

Zeta Potential: A positive or negative electrical charge that can build up around particles in a colloid.Zeta potential is the electrical potential that exists at the shear plane of a particle, which is some small distance from the surface.

Colloidal particles dispersed in a solution are

electrically charged due to their ionic characteristics and dipolar attributes. The development of a net charge at the particle surface affects the distribution of ions in the neighboring interfacial region, resulting in an increased concentration of counter ions (ion of charge opposite to that of the particles) close to the surface.

Each particle dispersed in a solution is surrounded by oppositely charged ions called fixed layer. Outside the fixed layer, there are varying compositions of ions of opposite polarities, forming a cloud-like area.

Thus an electrical double layer is formed in the region of the particle-liquid interface.

This double layer may be considered to consist of two parts: a inner region which includes ions bound relatively strongly to the surface and an outer, or diffuse region in which the ion distribution is determined by a balance of electrostatic forces and random thermal motion. The potential in this region, therefore, decays with the distance from the surface, until at a certance distance it becomes zero

When a voltage is applied to the solution in which particles are dispersed, particles are attracted to the electrode of the opposite polarity, accompanied by the fixed layer and part of the diffuse double layer. The potential at the boundary between this unit, that is to say at the above-mentioned shear plane between the particle with its ion atmosphere and the surrounding medium, is known as the Zeta Potential.

Zeta potential is a function of the surface charge

of a particle, any adsorbed layer at the interface and the nature and composition of the surrounding medium in which the particle is suspended. Zeta potential can be calculated with the following Smoluchowski's formula:

The principle of determining zeta potential is very simple. A controlled electric field is applied via electrodes immersed in a sample suspension and this causes the charged particles to move towards the electrode of opposite polarity. Viscous forces acting upon the moving particle tend to oppose this motion and an equilibrium is rapidly established between the effects of the electrostatic attraction and the viscosity drag. The particle therefore reach a constant terminal velocity.

Zettatechnology: In which zetta means 10^{21}, referring to the typical number of distinct designed parts in a product made by the systems we envision (molecular, mature, or molecular-manufacturing-based nanotechnology). The term refers to the implemented technology and its products, rather than to intermediate steps on the pathway.

ZIF (Zero-Inspection-Force) Socket: A socket in which package leads are readily accepted by the socket, and then firmly connected through cam action.

Zinc Finger: (1) Structural motif seen in many DNA-binding proteins, composed of a loop of polypeptide chain held in a hair-pin bend bound to a zinc atom. (2) A *zinc finger* is part of a protein that can bind to DNA. Zinc finger domains typically consist of two ß sheets, each carrying a cysteine residue, and an a helix carrying two histidine residues. The cysteine and histidine residues bind a zinc atom.

Zippocytes: In medical nanorobotics, a theorized medical nanorobot that can rapidly perform incision-wound repairs to the dermis and epidermis; dermal zippers.

Zone Refining: A metallurgical process for obtaining a highly pure metal that depends on continuously melting the impure material and recrystallizing the pure metal.

A purification technique for crystalline substances in which a heating system passes slowly over a bar of the material to be refined, creating a molten region that carries impurities with it across the bar. Also called *zone refining.*